NATO ASI Series

Advanced Science Institutes Series

A series presenting the results of activities sponsored by the NATO Science Committee, which aims at the dissemination of advanced scientific and technological knowledge, with a view to strengthening links between scientific communities.

The Series is published by an international board of publishers in conjunction with the NATO Scientific Affairs Division

A Life Sciences **B Physics**	Plenum Publishing Corporation London and New York
C Mathematical and Physical Sciences **D Behavioural and Social Sciences** **E Applied Sciences**	Kluwer Academic Publishers Dordrecht, Boston and London
F Computer and Systems Sciences **G Ecological Sciences** **H Cell Biology** **I Global Environmental Change**	Springer-Verlag Berlin Heidelberg New York London Paris Tokyo Hong Kong Barcelona Budapest

PARTNERSHIP SUB-SERIES

1. Disarmament Technologies	Kluwer Academic Publishers
2. Environment	Springer-Verlag
3. High Technology	Kluwer Academic Publishers
4. Science and Technology Policy	Kluwer Academic Publishers
5. Computer Networking	Kluwer Academic Publishers

The Partnership Sub-Series incorporates activities undertaken in collaboration with NATO's Cooperation Partners, the countries of the CIS and Central and Eastern Europe, in Priority Areas of concern to those countries.

NATO-PCO DATABASE

The electronic index to the NATO ASI Series provides full bibliographical references (with keywords and/or abstracts) to about 50000 contributions from international scientists published in all sections of the NATO ASI Series. Access to the NATO-PCO DATABASE compiled by the NATO Publication Coordination Office is possible in two ways:

- via online FILE 128 (NATO-PCO DATABASE) hosted by ESRIN, Via Galileo Galilei, I-00044 Frascati, Italy.
- via CD-ROM "NATO Science & Technology Disk" with user-friendly retrieval software in English, French and German (© WTV GmbH and DATAWARE Technologies Inc. 1992).

The CD-ROM can be ordered through any member of the Board of Publishers or through NATO-PCO, Overijse, Belgium.

Series H: Cell Biology, Vol. 95

Springer
Berlin
Heidelberg
New York
Barcelona
Budapest
Hong Kong
London
Milan
Paris
Santa Clara
Singapore
Tokyo

Flow and Image Cytometry

Edited by

Alain Jacquemin-Sablon

Centre National de la Recherche Scientifique
Laboratoire de Cytométrie (UPS 47)
7, rue Guy Moquet
F-94800 Villejuif, France

Springer

Published in cooperation with NATO Scientific Affairs Division

Proceedings of the NATO Advanced Study Institute "Progress in Flow and Image Cytometry", held at Villejuif, France, May 15–19, 1995

Library of Congress Cataloging-in-Publication Data

NATO Advanced Study Institute "Progress in Flow and Image Cytometry" (1995 : Villejuif, France)
Flow and image cytometry / edited by Alain Jacquemin-Sablon.
p. cm. -- (NATO ASI series. Series H, Cell biology ; vol. 95)
"Published in cooperation with NATO Scientific Affairs Division."
"Proceedings of the NATO Advanced Study Institute "Progress in Flow and Image Cytometry" held at Villejuif, France, May 15-19, 1995"--CIP t.p. verso.
Includes bibliographical references and index.

1. Flow cytometry--Congresses. 2. Imaging systems in biology--Congresses. I. Jacquemin-Sablon, Alain. II. Title. III. Series.
QH585.5.F56N38 1996
574.87'028--dc20 95-51230
CIP

ISBN-13: 978-3-642-64701-7 e-ISBN-13: 978-3-642-61115-5
DOI: 10.1007/978-3-642-61115-5

Softcover reprint of the hardcover 1st edition 1996

Typesetting: Camera ready by authors/editors
Printed on acid-free paper
SPIN 10477102 31/3137 - 5 4 3 2 1 0

FOREWORD

The second NATO-sponsored International Cytometry course, "Progress in Flow and Image Cytometry", was specifically structured to provide the most advanced techniques in flow and image cytometry currently available. The course provided two options for specialized training in the areas of (a) membrane dynamics and function (Option A) and (b) cell proliferation and gene expression (Option B). Twenty-four candidates were accepted for each course option which was composed of lecture and practical, "hands-on" laboratory exercices. The candidates were chosen on the basis of their interest and experience in the respective subject areas. A core curriculum consisting of topics on data management, control of cell proliferation, apoptosis, new probes in cytometry, cell sorting, and near field microscopy was presented to all participants.

Membrane transport and the mechanisms which control molecular traffic across plasma membranes are of considerable importance in attempts to modulate disease states and theses topics were addressed in Option A of the Course. Membrane pumping dynamics were included in the laboratory exercices as were topics on cell-cell interactions and T cell homing. Techniques using both flow and image cytometry for studying membrane transport and cell recognition are currently available. Advances in these areas have been made by the availability of new reagents that facilitate the speed and accuracy of the assays. Membrane fusion and fission in endocytosis and membrane fluidity and pinocytosis are the focus of many cytological investigations since the mechanisms involved are important to the understanding of many normal physiological processes as well as numerous disease conditions.

Recent developments in multiparameter flow cytometry (FCM) that provide for correlated analysis of a variety of cell kinetic parameters known to regulate and control cell proliferation were emphasized in Option B of the Course. Bromodeoxyuridine (BrdUrd) incorporation studies, for example, have significantly improved analysis of cell cycle progression, and the use of monoclonal antibodies in FCM immunofluorescence studies are providing new techniques for examining and quantitating proliferation markers (i.e., PCNA, Ki-67, p53 and cyclins) in different phases of the cell cycle. In addition, the role of apoptosis in regulation of heterogeneous cell types and in the induction of cell damage by various cycle perturbing agents is a primary aera of study today. Molecular cytogenetics as assayed both by flow and image cytometry have improved our understanding of chromatin organization and function. Studies on cell differentiation in normal cells are extremely important since determining the mechanisms involved could, alternatively, lead to a better understanding of how cells become transformed.

Data handling for analysis of univariate and multivariate distribution continues to be increasingly more complex, especially as the capabilities for assaying the numbers of variables increase with each new reagent or instrumentation feature that comes on the market. Cluster analysis, which attempts to analyze and correlate all the variables, is being explored in a number of laboratories. Data obtained from clinical samples stained with a battery of monoclonal antibodies serves as an excellent model for demonstrating the difficulties involved in such analysis. Coupled with the methods is the requirements for sorting and microscopic examination for precise identification of each subpopulation. Cell sorting features are available on many flow instruments but rarely do the users have the opportunity to obtain the proper instruction for sorting except in courses such as these. The fields of flow and image cytometry are advancing so quickly and the increase in the number of new techniques is so rapid only the continuation of these NATO sponsored courses can serve to provide the training necessary for the investigators to implement the most up-to-date technology in their research. Courses such as these, that include both lectures and laboratory experiences, are rare and expensive, and the support provided by NATO not only provides this unique educational opportunity but also represents a significant contribution to the international scientific community.

Co-Directors :

Dr. Alain Jacquemin-Sablon
CNRS, Laboratoire de Cytométrie, 94800 Villejuif France

Dr. Harry A Crissman
Los Alamos National Laboratory, Los Alamos, NM 87545 USA

Acknowledgements

Generous support from the following Institutions and Companies is gratefully acknowledged :

- The *Centre National de la Recherche Scientifique (CNRS)*, the *Association pour la Recherche sur le Cancer (ARC, Président, Jacques Crozemarie)*

- BECTON DICKINSON
- BIO-RAD
- CELL ROBOTICS
- COHERENT
- COULTRONICS
- MERIDIAN (D.G.L. Bioscience)
- ORTHO Diagnostic Systems
- PARTEC

We also wish to thank :

- Martine BARDISSA
- Claude BOUCHEIX
- Dominique LAIRD
- Linda PRITCHARD
- Arlette VERVISCH

for their expert assistance during the whole Course.

TABLE OF CONTENTS

PART I : MEMBRANE DYNAMICS AND FUNCTION

PART II : CELL PROLIFERATION AND GENE EXPRESSION

PART I

Membrane Dynamics and Function

ANALYSIS OF CD28 INTERACTIONS WITH ITS COGNATE COUNTER-RECEPTORS CD80 AND CD86.

Alemseged Truneh, Manjula Reddy, Patricia Ryan, Sally D.Lyn, Ilona Kariv, Christopher Eichman, Dominique Couez, Mark R.Hurle[1], Raffick P. Sekaly[3], Daniel Olive[3] and Raymond Sweet.

Department of Molecular Immunology
SmithKline Beecham Pharmaceuticals,
King of Prussia, Pennsylvania 19406, USA

Keywords :
CD28, CTLA-4, CD80, CD86, B7.1, B7.2, B70, IgSF, Ig-fold, Site Directed Mutagenesis, Receptor Recognition, Epitope Mapping.

Summary

CD28 serves as a co-signalling molecule for T cell activation through binding to its cognate counter-receptors CD80 and CD86, expressed on antigen presenting cells. This report summarizes studies conducted to determine the regions of CD28 which are involved in its interactions with CD80 and CD86, using site directed mutagenesis, CD28 mAb epitope mapping, receptor based adhesion assays and direct binding of Ig-fusion proteins to cell surface receptors. These studies

[1]*Department of Macromolecular Sciences, SmithKline Beecham Pharmaceuticals, King of Prussia, Pennsylvania 19406 USA.* [2]*IRCM, Montreal, Canada .* [3]*INSERM U119, Marseille, France*

NATO ASI Series, Vol. H 95
Flow and Image Cytometry
Edited by Alain Jacquemin-Sablon

show that, although the same overall region on CD28 may be involved in the interactions with CD80 and CD86, subtle but important differences distinguish recognition by the two molecules. These findings, along with other observations on the differential pattern of expression and tissue distribution of CD80 and CD86, and induction of differential cytokine profiles by the two counter-receptors, support the contention that these molecules play distinct roles in the regulation of immune responses *in vivo*.

Introduction

CD28 and the CD80 and CD86 Counter-receptors

The CD28 receptor, a member of the IgSF [1] , is a 44 kDa homodimer glycoprotein expressed on the surface of T-lineage cells including thymocytes and peripheral T cells in the spleen, lymph node and peripheral blood. Among human peripheral leucocytes, it is expressed by the majority of $CD4^+$ T cells (>85%) and a large proportion of $CD8^+$ T cells (~50%). CD28 interacts with its counter-receptors CD80 (originally referred to as B7 and more recently as B7.1) [2; 3] and CD86 (originally described as B70 or B7.2) [4; 5; 6], expressed on antigen presenting cells (APCs; see Figure 1), to deliver crucial co-stimulator signals for sustained activation of T cells, through its association via the cytoplasmic domain with PI3-kinase [7; 8]. Both CD80 [9] and CD86 [10; 5; 6] also recognize CTLA-4 [11] , a homolog of CD28, expressed transiently and at low receptor density on activated $CD8^+$ and $CD4^+$ T cells.

T Cell Interactions With Antigen Presesnting Cells

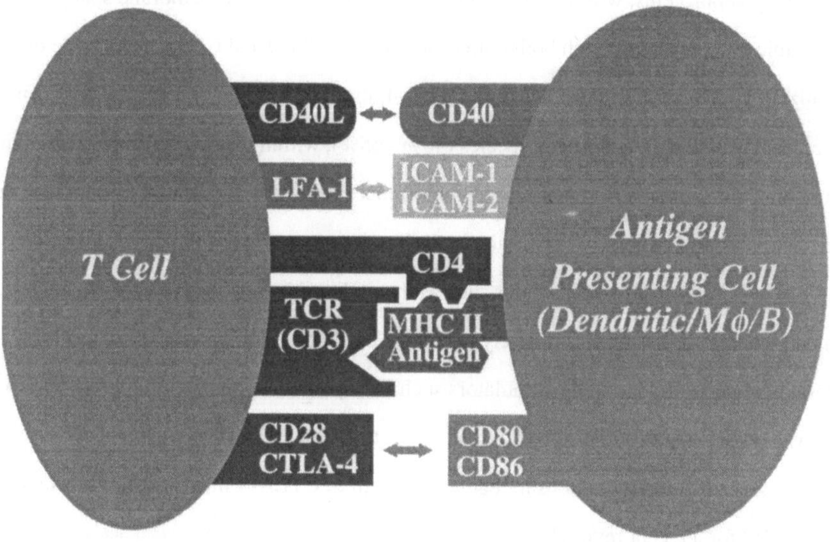

Figure 1

Ig Domain Structures of CD28, CTLA-4 and B7 (CD80)

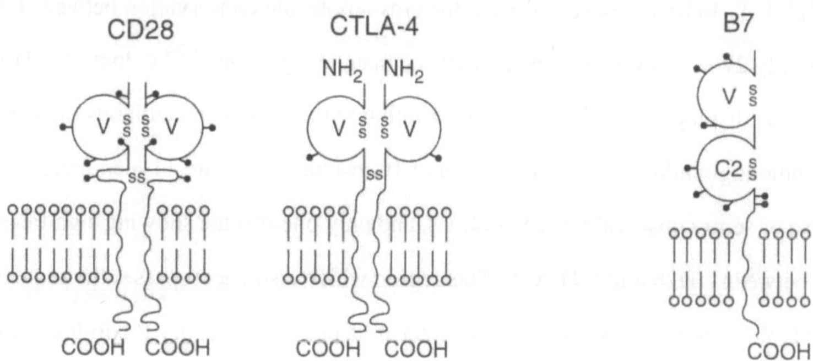

CD28 and its homolog CTLA-4 are homodimeric surface glycoproteins with a single IgV-like extracellular domain.
CD80 (B7.1) and its homolog CD86 (B7.2 or the N-terminal splice variant B70) are monomeric surface glycoproteins. A N-terminal IgV-like domain followed by an IgC-like domain comprise the extracellular region.

CD80 and CD86, which themselves are also members of the IgSF, are single chain, type I membrane glycoproteins, with IgV- and IgC-like domains and short membrane spanning and cytoplasmic domains. Although both human and murine CD80 and CD86 are capable of recognizing human and murine CD28 and CTLA-4, they show poor conservation between molecules (24% - 27%), and across species (46% - 65%), within the IgV- and IgC-like domains. Apart from their sequence divergence, other differences between CD80 and CD86 include their tissue distribution or pattern of expression in the basal state, and upregulation of the mRNA and protein in response to activation signals such as that induced by IFN-γ [10; 12; 5; 6; 13; 14]. For example, whereas CD86 is constitutively expressed on B cells and macrophages, CD80 expression requires induction by stimulators such IFN-γ. These observations and other *in vitro* and *in vivo* experiments with CTLA4-Ig and mAbs to CD80 and CD86 suggest that the two molecules serve different functions in regulating CD28 or CTLA-4-mediated T cell responses *in vivo* [see [15] for a recent review].

CD28 sequence conservation/divergence through speciation

Apart from human CD28 [16] and CTLA-4 [17], the mouse [18], rat [19], dog [20] and chicken [21] CD28 and mouse CTLA-4 [11; 22] have been cloned and show a reasonably high degree of cross-species conservation ranging from 50% - 92% amino acid identities for CD28 and 74% for CTLA-4. In both mouse and man, the cross-molecule conservation between CD28 and CTLA-4 is only 27% - 29% within the putative cognate recognition V-like domains. However, these molecules display a high degree of conservation within a region analogous to the CDR3 region of immunoglobulins where, in a stretch of 16 residues, there are 11 identities, 4 conservative or semi-conservative substitutions, and only one residue showing significant difference between CD28 and CTLA-4. This region which also encompasses the sequence " M Y P P P Y ", shows no cross-species or inter-molecule evolutionary flexibility, allowing not even conservative substitutions (see Figure 2).

Alignment of CD28 and CTLA-4 V-domains

```
                                                ◊
              <--A--->                  <---  B---->< CDR1<  ---C--->
                        10                   20         30           40
HuCD28      NKILVKQSPM  -LVAYDNAVNL  SCKYSYNLFS  REFRAS-LHKG
CaCD28      NKILVKQLPR  -LVVYNNEVNL  SCKYTYNLFS  KEFRAS-LYKG
MuCD28      NKILVKQSPL  -LVVDSNEVSL  SCRYSYNLLA  KEFRAS-LYKG
RaCD28      NKILVKQSPL  -LVVDNNEVSL  SCRYSYNLLA  KEFRAS-LYKG
ChCD28      NKILVAQRPL  -LIVANRTATL  VCNYTYNGTG  KEFRAS-LHKG
HuCTLA-4    KAMHVAQPAV  VLASSRGIASF  VCEYASPGKA  TEVRVTVLRQA
MuCTLA-4    EAIQVTQPSV  VLASSHGVASF  PCEYSPSHNT  DEVRVTVLRQT

                   «                              »
              <--C'--><--  CDR2-   -C"-   ---   <-D--><   -DE--><----E---
                      50           60           70           80
HuCD28      LDSAV-EVCVV  YGNYSQQLQV  YSKTGFNCDG  KL-GNESVTFY
CaCD28      VDSAV-EVCVV  NGNYSHQPQF  YSSTGFDCDG  KL-GNETVTFY
MuCD28      VNSDV-EVCVG  NGNFTYQPQF  RSNAEFNCDG  DF-DNETVTFR
RaCD28      VNSDV-EVCVG  NGNFTYQPQF  RPNVGFNCDG  NF-DNETVTFR
ChCD28      TDSAV-EVCFI  SWNMT-KINS  NSNKEFNCRG  IH-DKDKVIFN
HuCTLA-4    -DSQVTEVCAA  T--YMMGNEL  TFLDDSICTG  TSSGNQ-VNLT
MuCTLA-4    -NDQMTEVCAT  T--FTEDNTV  GFLDYPSCSG  TFNESR-VNLT

                            ◊                                          ^
              -->       <-  --F---><-----  CDR3---><---  ---G------->
                      90          100          110          120
HuCD28      LQNLYVNQTD  IYFCKIEVMY  PPPYLDNEKS  NGTIIHVKGK  HLCPSPL
CaCD28      LRNLFVNQTD  IYFCKIEVMY  PPPYIGNEKS  NGTIIHVKEK  HLCPDEL
MuCD28      LWNLHVNHTD  IYFCKIEFMY  PPPYLDNERS  NGTIIHIKEK  HLCHTQS
RaCD28      LWNLDVNHTD  IYFCKIEVMY  PPPYLDNEKS  NGTIIHIKEK  HLCHAQT
ChCD28      LWNMSASQTD  IYFCKIEAMY  PPPYVYNEKS  NGTVIHVRET  PIQTQEP
HuCTLA-4    IQGLRAMDTG  LYICKVELMY  PPPYYLGIG-  NGAQIYVIDP  EPCPDSD
MuCTLA-4    IQGLRAVDTG  LYLCKVELMY  PPPYFVGMG-  NGTQIYVIDP  EPCPDSD
```

Figure 2

Alignment of the V-like domains of CD28 and CTLA-4 from different species. This alignment was based on the experimentally determined N-terminus for human CD28 and assumes similar N-termini for other species. Those for CTLA-4 are as predicted from the cDNA sequences. The strand and loop assignments are based on IgV domain consensus sequences (not shown).

Results

Alignment of CD28 and CTLA-4 with IgV domains:

To map the Ig-fold pattern of human CD28, CD28 and CTLA-4 sequences from the different species were aligned with consensus IgV region sequences (Kabat et al. 1991). Based on this alignment, the β-strands (β-sheets) of an Ig-fold molecule and putative exposed loops, including CDR-like regions, were identified (Figure 2). This alignment shows 9 β-strands with 3 CDR-like regions and one non-CDR-loop (DE-loop) in the membrane distal regions of the molecule. This alignment also places the fully conserved MYPPPY sequence, which had originally been ascribed to the stalk region (i.e. between the IgV and transmembrane regions) of these molecules [22], within the CDR-3-like region. It is to be noted that CDR3 regions on antibody structures form a major component of the antigen recognition site [reviewed in [23]].

Generation of CD28 mutants and evaluation of binding to mAbs, CD80 and CD86

A number of single and multiple residue substitution or truncation mutants were generated within the IgV-like domain of CD28 and expressed on the surface of L cells or as functionally active receptor on the surface murine T cell hybridoma lines [24]. CD80 and CD86 were similarly expressed as cell surface proteins in L cells. All cells were enriched for homogneous high receptor density expression by 2 sort cycles by flow cytometry, using universally reactive mAbs as probes. CD28, CD80 and CD86 were also expressed as soluble, secreted Ig-fusion proteins in CHO cells.

Reactivity of mutants or wild type CD28 with mAbs, CD80 and CD86 was evaluated in several ways: 1) in cell-cell adhesion assays, 2) cell-protein adhesion assays using the Ig-fusion proteins, and 3) by direct binding of mAbs or Ig-fusion proteins to cell surface expressed CD28, by flow cytometry [25].

Epitope map of CD28 mAbs

A panel of previously described CD28 mAbs was used to assess the structural integrity of the mutants. The cross-blocking patterns of these mAbs are shown in Table 1. This panel cluster into at least 6 groups according to their cross-blocking characteristics [26], their ability to stimulate

CD28-mediated IL-2 production, or to block CD28 and CD80 dependent cell-cell adhesion [27]. All these mAbs, with the exception of CD28.4, block the interaction of CD28 with CD80 as determined by an adhesion assay based on CD28 and CD80 transfected cells [27] and data not shown]. This panel of CD28 mAbs was used to assess the structural integrity of all CD28 mutants in our studies.

Table 2 summarizes the binding characteristics of CD28 mAbs to the CD28 mutants. Consistent with the previous characterizations, the mAbs cluster into different groups according to the their pattern of binding to the CD28 mutants. Apart from 3 cell lines transfected with the M1, M6 and M14 mutants of CD28, all other mutants bound one or more mAbs, indicating that the mutations did not cause severe structural distortions. The absence of binding of all the mAbs to the M1, M6 and M14 mutants was interpreted as an indication that these mutations caused severe structural deformations of the protein resulting in loss of native epitopes, and/or lack of surface expression. These mutants were excluded from further analysis. All other mutants were analysed for interactions with CD80 or CD86 using cell-cell or cell-protein adhesion assays or for direct binding to CD80-Ig by flow cytometry.

In analysing the epitope map generated by cross-blocking studies in conjunction with that generated by mutant analysis, a cluster which includes CD28.1, CD28.2 and CD28.3, which had previously been grouped together based on cross-blocking and functional studies [26; 27], was affected by the same set of mutants, corroborating the earlier data. CD28.5, which appears to cause IL-2 production in CD28-transfected T cell hybridomas, without the need for cross-linking either by ligand immobilization or with secondary antibodies, displayed a distinct pattern in the mutant binding assay. One mAb, CD28.6 was uniquely unaffected by any of the mutants directed at the membrane distal CDR-like regions or the DE-loop, indicating that this mAb may bind to the β-strand / β-sheet structure within the V-like domain of CD28. Since this mAb, nevertheless, still blocks CD28 binding to CD80 or CD86 (ref. [27] and unpublished), its epitope must lie in a region proximal to the cognate recognition site.

Analysis of CD80 and CD86 binding to CD28 mutants

CD80 and CD86 were evaluated for their ability to binding CD28 mutants in several different binding assays, including adhesion of CD80 and CD86 exprssing cells to cells expressing wild type or mutant CD28, direct binding of CD80-Ig or CD86-Ig to CD28 expressing cells in adhesion assay format or by flow cytometry. A summary of these results, represented in Table 3, shows that truncation of the "M Y P P P Y " sequence (M8), or its substitution with the flexible [28] linker "S G G G " (M13), caused loss of reactivity with CD80 expressing cells. Similar loss of reactivity was observed for mutants which contained replacements of the internal proline residues within this motif (M3 and M7). Loss of reactivity was also observed when the second flanking tyrosine and two downstream residues were changed from " Y L D " to " L A S " (M4). A conservative substitution of the two flanking tyrosine residues with phenylalanine (M11) had no effect on binding to CD80. All mutations which caused loss of reactivity with CD80 also had the same effect on reactivity with CD86. However, in contrast to its ability to bind CD80, M11 showed complete loss of reactivity with CD86.

Localization of CD28 Residues Involved in CD80 and CD86 Recognition on an Ig-fold Structure

To localize the disposition of putative cognate-receptor recognition elements on the CD28 molecule, the CD28 sequence alignment in Figure 2 was used to model the folding patterns of CD28 based on known IgV structures. The model (Figure 3) shows 9 β-strands which form the characteristic two β-sheets of all Ig-fold structures [29; 23] . Unlike IgV structures, the CD28 structure appears to be stabilized by two pairs of cysteines (rather than one pair), which are predicted to form inter-sheet disulfide bonds. It is noteworthy that, M14, a mutant which includes a cysteine to alanine switch of one of this putative novel C'-D pair, caused disruption of receptor expression, presumably as a result of a loss of structural integrity of the V domain.

According to this model, a number of residues or stretches of residues which are highly conserved between species, or between CD28 and CTLA-4 fall within the 9 β-strands or their boundary regions, implying a contribution towards the tertiary or quaternary sturctures of these proteins. Notably though, the region containing the hallmark residues "MYPPPY", localizes to the

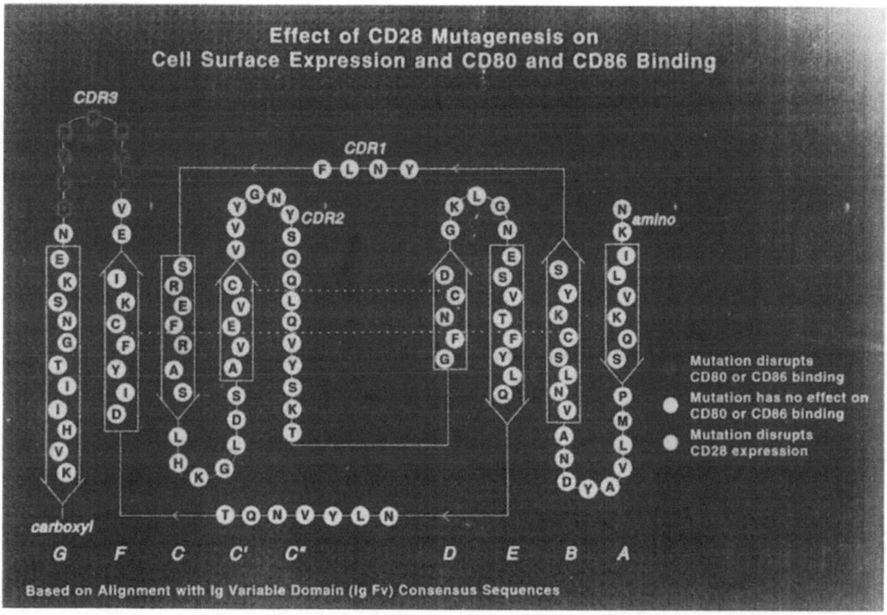

Figure 3

The V-like domain of CD28 is represented as an Ig-fold structure with 9 β-strands (see *italics* lettering below structure), 3 CDR-like loops and one non-CDR (DE) loop. Single or multiple residue mutations which caused loss of CD80 or CD86 reactivity or loss of expression, and mutations which had no effect on binding or expression are represented.

CDR3-like loop, which in antibody structures contributes significantly to antigen recognition. As described above, truncation of this sequence or its substitution by a flexible stretch of irrelevant residues caused total loss of counter-receptor binding, without causing generalized disruption of the CD28 structure.

Mutations which disrupted CD28 expression, (M1, M6 & M14) localized to β-strands. Another mutant (M5), which localized within a putative β-strand, did not disrupt CD28 expression, but also did not affect recognition of cognate counter-receptors. This mutant localizes to the G-strand, which is at the end of one of the β-sheet structures, and is more exposed on IgV structures than the other strands [23]. Other mutants which did not cause loss of expression or receptor binding localized within, or at, the boundaries of the CDR1-like, CDR2-like or DE loops.

Discussion and Conclusions

The most distinguishing feature of the CD28 and CTLA-4 molecules is the conservation through speciation of the highly unusual sequence " M Y P P P Y " (residues 99-104 on CD28), which on an Ig-fold structure localizes to a CDR3-like region. Other highly conserved residues within this region, such as residues 89-96 appear to fall within β-strand / β-sheet structures of an Ig fold, and thus may contribute more towards the tertiary or quaternary structures of these molecules than to ligand recognition. For example, the fully conserved Tyr92 and its flanking residues, which were localized to the F strand of CD28, represent fully or partially buried, solvent inaccessible side chains on Ig light chain structures [23].

Although fully conserved in the known CD28 and CTLA-4 molecules thus far identified [16; 11; 18; 22; 19; 20; 21], " M Y P P P Y " does not exist in any other reading frame in the GENBANK / EMBL databases, although " Y P P P Y " was found in the reading frame of approximately 60 other molecules. In other proteins whose structures have been determined, the highly constrained nature of proline generally favours turns and nonrepetitive structures at the end of strands and helices [28]. Two prolines in a row favour extended conformations such as proline rich stretches in Ig hinge regions. The complete conservation of this highly unusual motif in all

CD28 and CTLA-4 genes thus far cloned, and its localization to a CDR-like region, strongly suggest a role in receptor recognition. Truncation of this region caused loss of binding to CD80 and CD86 without loss of surface expression, and with the retention of some antibody epitopes. The latter observations suggest that this truncation did not result in gross overall deformation of the CD28 structure. Similar results were also obtained when a glycine-based flexible hinge was subsituted for the original sequence, to bridge the gap between the two strands. Other mutations in the proximity of, or within, the CDR1-like, CDR2-like or DE loops in the membrane distal region of the V loop had very little effect on cognate interaction. None of the mutations in this region decreased receptor expression or affected all mAb epitopes. In sharp contrast, mutations introduced within putative β-strands on this molecule (M1 and M6 localizing to β-strand C, and M14 localizing to β-strand D), resulted in the complete loss of expression, presumably due to disruption of the β-sheet structures. Although, in general, mutations within β-strands appear to be poorly tolerated for surface expression, a 3 residue non-conservative substitution within the G-strand (EKS 108-110 SNA), caused neither loss of surface expression nor disruption of CD80 reactivity. This mutation also did not disrupt any of the CD28 mAb epitopes. G-strands on IgV_L domains or on the CD8α homodimer structure [30], at the end of one of the β-sheets of a V-domain, form part of the dimer interface with the C' strand of an aposing V domain. All of these observations are in line with the working Ig-fold model of CD28 presented in Figure 8.

In a concurrent study, Peach et al [31] have also implicated the " M Y P P P Y " sequence for binding of CD28 and CTLA-4 to CD80. Through homolog substitutions between CD28 and CTLA-4, that study also suggested that the CDR1-like region may contribute towards the binding of CD28 to CD80. In the current study, a homolog switch within this region, that of a murine sequence into the human structure (oringinally designed for purposes of mAb epitope mapping), had very little effect on CD28 binding. More precise determination of the nature and extent of participation of the CDR1-like region of CD28 in cognate recognition will require further analysis, including single residue saturation mutagenesis.

A conservative substitution of the flanking two Tyr residues within the " M Y P P P Y " motif to Phe resulted in the total loss of binding to CD86 with little or no effect on CD80 recognition. To date, no differences in binding to CD28 have been reported between CD80 and CD86. Indeed, the Ig-fusion derivatives of these receptors show similar binding characteristics with respect to CD28 and CTLA-4, when measured by adhesion assays or direct binding by flow cytometry (unpublished). Subtle differences in affinities or binding energetics await application of more refined biophysical techniques such as analytical ultracentrifugation, plasmon resonance and/or microcalorimetry. What is apparent from sequence analysis is that, although CD80 and CD86 are poorly conserved within the putative CD28 binding V-like domain, they share some local spots of homology. One particular region, which displays 3 identities and 3 conservative substitutions within a stretch of 9 residues (residues 83-92 in CD80: "IWPEYKNRT"), appears to be involved in CD28 and CTLA-4 recognition [32]. Within this region, the tryptophan on CD80 implicated in CD28 recognition, has been substituted for with a histidine on CD86. It is interesting to speculate whether this conservative substitution contributes to the subtle differences in the interactions of these molecules, observed in the current study. These differences in the recognition patterns between CD80 and CD86 for the CD28 counter-receptor add to the accumulating evidence that CD80 and CD86 serve distinct functions in the regulation of the immune system (reviewed in [15]).

Acknowledgements

We would like to thank P. Golstein for providing the human CTLA-4 cDNA, M. Chaikin for construction and expression of CD80-Ig, J. Culp and B. Hellmig for purification and D. McNulty for N-terminal sequencing of the Ig-fusion proteins.

References

1. Williams, A. F., & Barclay, A. N. (1988). *The immunoglobulin superfamily--domains for cell surface recognition. [Review].* Annual Review of Immunology, 6: 381-405.
2. Linsley, P. S., Clark, E. A., & Ledbetter, J. A. (1990). *T-cell antigen CD28 mediates adhesion with B cells by interacting with activation antigen B7/BB-1.* Proc Natl Acad Sci U S A, 87: 5031-5.
3. Freeman, G. J., Gray, G. S., Gimmi, C. D., Lombard, D. B., Zhou, L. J., White, M., Fingeroth, J. D., Gribben, J. G., & Nadler, L. M. (1991). *Structure, expression, and T cell costimulatory activity of the murine homologue of the human B lymphocyte activation antigen B7.* Journal of Experimental Medicine, 174: 625-31.
4. Azuma, M., Cayabyab, M., Buck, D., Phillips, J. H., & Lanier, L. L. (1992). *CD28 interaction with B7 costimulates primary allogeneic proliferative responses and cytotoxicity mediated by small, resting T lymphocytes.* J Exp Med, 175: 353-60.
5. Freeman, G. J., Borriello, F., Hodes, R. J., Reiser, H., Gribben, J. G., Ng, J. W., Kim, J., Goldberg, J. M., Hathcock, K., Laszlo, G., & et, a. l. (1993). *Murine B7-2, an alternative CTLA4 counter-receptor that costimulates T cell proliferation and interleukin 2 production.* J Exp Med, 178: 2185-92.
6. Freeman, G. J., Gribben, J. G., Boussiotis, V. A., Ng, J. W., Restivo, V. A. J., Lombard, L. A., Gray, G. S., & Nadler, L. M. (1993). *Cloning of B7-2: a CTLA-4 counter-receptor that costimulates human T cell proliferation [see comments].* Science, 262: 909-11.
7. Pages, F., Ragueneau, M., Rottapel, R., Truneh, A., Nunes, J., Imbert, J., & Olive, D. (1994). *Binding of phosphatidylinositol-3-OH kinase to CD28 is required for T-cell signalling.* Nature, 369: 327-9.
8. Stein, P. H., Fraser, J. D., & Weiss, A. (1994). *The cytoplasmic domain of CD28 is both necessary and sufficient for costimulation of interleukin-2 secretion and association with phosphatidylinositol 3'-kinase.* Molecular & Cellular Biology, 14: 3392-402.
9. Linsley, P. S., Brady, W., Urnes, M., Grosmaire, L. S., Damle, N. K., & Ledbetter, J. A. (1991). *CTLA-4 is a second receptor for the B cell activation antigen B7.* J Exp Med, 174: 561-9.
10. Azuma, M., Ito, D., Yagita, H., Okumura, K., Phillips, J. H., Lanier, L. L., & Somoza, C. (1993). *B70 antigen is a second ligand for CTLA-4 and CD28.* Nature, 366: 76-9.
11. Brunet, J. F., Denizot, F., Luciani, M. F., Roux, D. M., Suzan, M., Mattei, M. G., & Golstein, P. (1987). *A new member of the immunoglobulin superfamily--CTLA-4.* Nature, 328: 267-70.
12. Boussiotis, V. A., Freeman, G. J., Gribben, J. G., Daley, J., Gray, G., & Nadler, L. M. (1993). *Activated human B lymphocytes express three CTLA-4 counterreceptors that costimulate T-cell activation.* Proc Natl Acad Sci U S A, 90: 11059-63.
13. Hathcock, K. S., Laszlo, G., Dickler, H. B., Bradshaw, J., Linsley, P., & Hodes, R. J. (1993). *Identification of an alternative CTLA-4 ligand costimulatory for T cell activation [see comments].* Science, 262: 905-7.
14. Larsen, C. P., Ritchie, S. C., Hendrix, R., Linsley, P. S., Hathcock, K. S., Hodes, R. J., Lowry, R. P., & Pearson, T. C. (1994). *Regulation of immunostimulatory function and costimulatory molecule (B7-1 and B7-2) expression on murine dendritic cells.* Journal of Immunology, 152: 5208-19.
15. Guinan, E. C., Gribben, J. G., Boussiotis, V. A., Freeman, G. J., & Nadler, L. M. (1994). *Pivotal role of the B7:CD28 pathway in transplantation tolerance and tumor immunity.* Blood, 84: 3261-3282.

16. Aruffo, A., & Seed, B. (1987). *Molecular cloning of a CD28 cDNA by a high-efficiency COS cell expression system.* Proc Natl Acad Sci U S A, 84: 8573-7.

17. Dariavach, P., Mattei, M. G., Golstein, P., & Lefranc, M. P. (1988). *Human Ig superfamily CTLA-4 gene: chromosomal localization and identity of protein sequence between murine and human CTLA-4 cytoplasmic domains.* European Journal of Immunology, 18: 1901-5.

18. Gross, J. A., St, John, T., & Allison, J. P. (1990). *The murine homologue of the T lymphocyte antigen CD28. Molecular cloning and cell surface expression.* Journal of Immunology, 144: 3201-10.

19. Clark, G. J., & Dallman, M. J. (1992). *Identification of a cDNA encoding the rat CD28 homologue.* Immunogenetics, 35: 54-7.

20. Pastori, R. L., Milde, K. F., & Alejandro, R. (1994). *Molecular cloning of the dog homologue of the lymphocyte antigen CD28.* Immunogenetics, 39.

21. Young, J. R., Davison, T. F., Tregaskes, C. A., Rennie, M. C., & Vainio, O. (1994). *Monomeric homologue of mammalian CD28 is expressed on chicken T cells.* Journal of Immunology, 152: 3848-51.

22. Harper, K., Balzano, C., Rouvier, E., Mattei, M. G., Luciani, M. F., & Golstein, P. (1991). *CTLA-4 and CD28 activated lymphocyte molecules are closely related in both mouse and human as to sequence, message expression, gene structure, and chromosomal location.* J Immunol, 147: 1037-44.

23. Padlan, E. A. (1994). *Anatomy of the antibody molecule. [Review].* Molecular Immunology, 31: 169-217.

24. Couez, D., Pages, F., Ragueneau, M., Nunes, J., Klasen, S., Mawas, C., Truneh, A., & Olive, D. (1994). *Functional expression of human CD28 in murine T cell hybridomas.* Molecular Immunology, 31: 47-57.

25. Truneh, A., Reddy, M., Ryan, P., Lyn, S. D., Eichman, C., Couez, D., Hurle, M. R., Sekaly, R. P., Olive, D., & Sweet, R. (1995). *Differential recognition by CD28 of its cognate counter-receptors CD80 (B7.1) and B70 (B7.2): Analysis by site directed-mutagenesis.* Molecular Immunology, in press.

26. Nunes, J., Klasen, S., Ragueneau, M., Pavon, C., Couez, D., Mawas, C., Bagnasco, M., & Olive, D. (1993). *CD28 mAbs with distinct binding properties differ in their ability to induce T cell activation: analysis of early and late activation events.* Int Immunol, 5: 311-5.

27. Truneh, A., Reddy, M., Couez, D., Kokolis, C., Eichman, C., Fargeas, C., Sekaly, R., Sweet, R., & Olive, D. (1994). *Analysis of CD28- and B7-mediated T cell adhesion and IL-2 production using a panel of mAbs and CTLA4-Ig and B7-Ig.* Leukocyte Typing V, in press.

28. Richardson, J. S., & Richardson, D. C. (1989). *Principles and patterns of protein conformation.* In G. D. Fasman (Eds.), Prediction of protein structure and the principles of protein conformation. (pp. p. 1-98.). New York: Plenum Press.

29. Amzel, L. M., & Poljak, R. J. (1979). *Three-dimensional structure of immunoglobulins. [Review].* Annual Review of Biochemistry, 48: 961-97.

30. Leahy, D. J., Axel, R., & Hendrickson, W. A. (1992). *Crystal structure of a soluble form of the human T cell coreceptor CD8 at 2.6 A resolution.* Cell, 68: 1145-62.

31. Peach, R. J., Bajorath, J., Brady, W., Leytze, G., Greene, J., Naemura, J., & Linsley, P. S. (1994). *Complementarity determining region 1 (CDR1)- and CDR3-analogous regions in CTLA-4 and CD28 determine the binding to B7-1.* J Exp Med, 180: 2049-58.

32. Fargeas, C. A., Truneh, A., Reddy, M., Hurle, M., Sweet, R., & Sekaly, R.-P. (1995). *Identification of residues in the V domain of CD80 (B7-1) implicated in functional interactions with CD28 and CTLA-4.* Journal of Experimental Medicicine, in press.

Cross-blocking pattern of CD28 mAbs used in current study

	CD28.1	CD28.2	CD28.3	CD28.4	CD28.5	CD28.6	248	L293
CD28.1	++	++	++	-	++	-	-	nd
CD28.2	++	++	++	-	++	++	++	nd
CD28.3	++	++	++	-	++	-	-	nd
CD28.4	-	-	-	nd	nd	-	nd	nd
CD28.5	++	++	++	-	++	-	-	nd
CD28.6	-	++	-	-	-	++	++	nd
248	-	++	-	nd	++	++	nd	nd
L293	nd	nd	nd	nd	nd	nd	nd	nd

Table 1

mAb Binding to CD28 Mutants

MT #	Mutation	CDR	CD28.1	CD28.2	CD28.3	CD28.4	CD28.5	CD28.6	L293	248
	mAb Subgroup ->		1.1.1	1.2.2	1.1.1	2.3.6	1.2.1	1.3.4	1.2.	1.1.5
	wild type		++	++	++	++	++	++	++	++
M10	FS (29-30) LA	1	++	++	++	+	++	++	++	++
M1	REFR (31-34) NSAN	β-strand C	-	-	-	-	-	-	-	-
M6	EFR (32-34) NFE	β-strand C	-	-	-	-	-	-	-	-
M12	YSQ (54-56) FTY	2	++	++	++	++	++	++	++	++
M2	SQQ (55-57) ATS	2	++	+	++	++	++	++	++	++
M14	NCD (67-69) SAT	β-strand D	-	-	-	-	-	-	-	-
M9	KLG (71-73) DFD	DE loop	++	++	++	++	++	++	++	++
M8	MYPPPY del	3	-	-	-	-	-	++	-	-
M13	MYPPPY to SGGG	3	-	-	-	-	-	++	-	-
M3	Y100L,P102G	3	-	-	-	-	-	++	-	++
M11	Y100F, Y104F	3	++	++	+	++	++	++	++	++
M7	P101G, P103L	3	-	-	-	-	-	++	-	+
M4	YLD (104-106) LAS	3	++	++	++	++	-	++	+-	++
M5	EKS (108-110) SNA	β-strand G	++	++	++	++	++	++	+-	++

A panel of CD28 mAbs was used to assess expression of the CD28 mutants. Consistent with previous mAb competition studies, the epitopes for these mAbs cluster into several groups based on their recognition of the HuCD28 mutants. The absence of binding for M1, M6, and M14 indicates a lack of expression of these mutants on the cell surface. All other CD28 mutants were recognized by at least 1 mAb. With the above exceptions, none of the mutants appeared to affect the epitope for mAb 28.6 and this antibody was use for relative quantitation of the mutants.

Footnote: Binding of mAbs to wild type or mutant CD28 expressing cells was determined by flow cytometry with using FITC-labelled goat-anti-mouse antibody as a secondary reagent. Mean fluorescence values are represented as "++" = strong binding, "+" = moderate binding and "-" = no binding.

Table 2

Summary of CD28 Mutant Expression and Binding to CD80 and B70

MT #	Mutation	CDR	Expression	CD80	B70
	wild type		+	+	+
M10	FS (29-30) LA	1	+	+	+
M1	REFR (31-34) NSAN	β-strand C	-	ND	ND
M6	EFR (32-34) NFE	β-strand C	-	ND	ND
M12	YSQ (54-56) FTY	2	+	+	+
M2	SQQ (55-57) ATS	2	+	+	+
M14	NCD (67-69) SAT	β-strand D	-	ND	ND
M9	KLG (71-73) DFD	DE loop	+	+	+
M8	MYPPPY del	3	+	-	-
M13	MYPPPY to SGGG	3	+	-	-
M3	Y100L,P102G	3	+	-	-
M11	Y100F, Y104F	3	+	+	-
M7	P101G, P103L	3	+	-	-
M4	YLD (104-106) LAS	3	+	-	-
M5	EKS (108-110) SNA	β-strand G	+	+	+

Expression data for this table ("+" = good expression, "-" = no expression), was obtained from flow cytometry studies using mAbs CD28.6 and 248. Data for CD80 and B70 binding ("+" = good binding, "-" = no binding), was obtained from adhesion assays and from flow cytometry studies with of Ig-fusion proteins.

Table 3

MEMBRANE TRANSPORT DYNAMICS

James V.Watson and Caroline Dive[1]
MRC Clinical Oncology Unit, The Medical School
Hills Road
Cambridge
United Kingdom CB2 2QH

Membrane transport and the mechanisms which control molecular traffic across the plasma membrane are of considerable importance in attempts to modulate disease states. This is a very important area to which flow cytometry is well suited but this has not yet been afforded the attention it either should or could have received. Transport of any agent across the external cell membrane occurs by at least three mechanisms namely, passive diffusion, facilitated diffusion and active transport. In passive diffusion the rate of increase of the intracellular concentration is directly proportional to the concentration gradient. With facilitated diffusion the traffic involves specific interactions between transport molecules and ligand but, unlike passive diffusion, it is temperature sensitive and exhibits saturation kinetics. However, movement against a concentration does not take place. Active transport mechanisms are able to concentrate ligand against a gradient using energy dependent systems. We tend to regard transport as a one way system from without to within but clearly, the mechanisms also apply to transport from inside the cell to outside, which has important implications in drug resistance mechanisms.

[1] Departement of Molecular Pharmacology, Stopford Building, Department of Physiological Science, University of Manchester, Oxford Road - Manchester, United Kingdom M13 9PT

NATO ASI Series, Vol. H 95
Flow and Image Cytometry
Edited by Alain Jacquemin-Sablon

Flow cytometric quantitation depends on fluorescence and any measurement of ligand transport must rely on at least one of a number of principles. First, the ligand itself may be fluorescent and its uptake and/or loss from the cell population is followed versus time. Second, a fluorescent reporter molecule may be available whose transport minics that of the ligand of interest. Third, quenching of a reporter molecule by the ligand of interest may take place which can be followed temporally. Fourth, fluorescent analogues may be available and finally, indirect measurements of intracellular drug content can be made by using inhibition of enzyme kinetic.

Measurement of any dynamic processes necessarily requires that time is included as a parameter and as this is the most important single measurement we will start with methods to incorporate it in the data base.

1) Incorporation of time

A number of different techniques have been introduced to incorporate time into the data base. Watson et al (1977) used a continuous interrupted sampling technique where the cells were continuously flowing through the instrument. The data acquisition computer was instructed to record for 5 seconds then wait for 10 seconds before recommencing acquisition. During the interval between recordings the median of the population was calculated and printed out. This method had resolution problems, particularly for "fast" reactions, as the fluorescence from the population was increasing during the data acqusition interval hence the medians tended to be overestimated. Martin and Swartzenruber (1980) overcame this very elegantly by incorporating time directly into the data base. Their initial method used a voltage ramp generator where the potential increased linearly with time and a recording of the voltage was made from the ramp generator each time a cell was analysed in the flow chamber. This method had one disadvantage, namely that time could only be recorded over an interval of about 15 minutes. However, immediately following the introduction of continouus time recording by Martin and Swarzendruber (1980) we incorporated the time stamp from the data acquisition computer directly into the data base. Hence, events were timed automatically with 50ms resolution by the computer clock and this time, in the form of the number of clock "ticks" (one tick per 50ms), was recorded for each event analysed.

Keij et al (1989) have incorporated "pseudo" time into list-mode data by inference from the computer recorded duration of the run. The latter was divided sequentially into 255 equal channels which produced a time-related factor proportional to time. This is not the ideal method for incorporating time into the data base as extreme care must be taken to keep the flow rate constant. However, it is the only compromise possible for commercial instruments which do not include the facility to incorporate time in the data base from the computer clock.

2) Fluorescent ligands

The commonly used anthracyclines, daunorubicin and adriamycin, are both fluorescent and can be used directly to study cellular accumulation and was first demonstrated by Krishan and Ganapathi (1980). Durand and Olive (1981) used similar direct fluorescence techniques to quantitate cellular adriamycin and additionally they showed time courses for uptake and compared nuclei with whole cells. Some of these data are reproduced in figure 1.

Multiparameter analysis of adriamycin and daunomycin uptake in mouse bone marrow and acute myeloid leukaemia cells has been carried out by Sonneveld and van den Engh (1981) using forward and 90° light scatter properties to identify different subsets (Visser, van den Engh and van Bekkum, 1980). Fluorescence was related directly to cytotoxicity and myeloid progenitor cells were an order of magnitude more sensitive to daunomycin compared with adriamycin, a result which correlated with fluorescence.

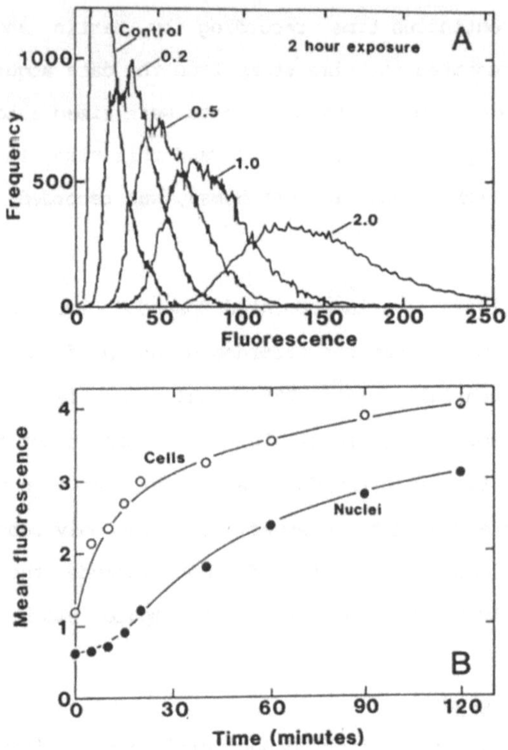

FIGURE 1) Panel A shows adriamycin fluorescence distributions for 2 hour exposures at the concentrations, in $\mu g\ M^{-1}$, indicated adjacent to each histogram. Time courses for the increase in adriamycin fluorescence for whole cells (O) and nuclei (●) exposed to 1.0 $\mu g\ ml^{-1}$ are shown in panel B. Redrawn from Durand and Olive (1981).

In contrast, the fluorescence from acute myeloid leukaemia cells was greater for adriamycin than for daunomycin. Further studies of daunomycin fluorescence uptake by rat bone marrow were carried out by Nooter, van den Engh and Sonneveld (1983) where dead cells, lymohocytes, blast cells, granulocytes and red cells were separately identified on light scatter. At higher concentrations the lymphocyte fraction exhibited greater fluorescence than granuloctyes and at given concentrations the dead cells exhibited considerably greater fluorescence than viable cells. These various effects were attributed to quantitative differences in active transport of these agents into different cell types.

Anthracyclines are actively transported both into and out of cells, but by different pump mechanisms. One of the efflux pumps is the 170 kd membrame associated glycoprotein (pg^{170}) encoded by the multi-drug resistance gene, MDR-1. Extensive evidence now exists which shows that multi-drug resistant cells have increased numbers of membrane associated gp^{170} molecules and as a result MDR cells exhibit lower fluorescence on exposure to anthracyclines compared with sensitive cells as these agents are actively transported out of MDR cells with greater efficiency. Ross et al (1988) have used this property to cell sort and isolate highly MDR cells at a very low frequency from within a sensitive P388 cell population on duanorubicin fluorescence. The correct functioning of the efflux pump mechanism can be modulated by Ca^{2+} channel blockers (chlorpromazine, trifluoperazine and verapamil, Tsuro et al, 1982). A number of studies on the modulation of anthracycline uptake by Ca^{2+} channel blockers have been carried out using flow cytometry in resistant and sensitive cells in both animal and human systems (Krishan et al, 1986, 1987; Ross et al, 1986; Nooter et al, 1989; Morgan et al, 1989; Watson, Morgan and Dive, 1989; Watson, Morgan and Smith, 1989). The "bottom line" in all these studies is that the lower accumulation of anthracyclines in multi-drug resistant cells can be partially reversed by exposure to Ca^{2+} channel blockers which interfere with the efflux pump mechanism.

Some of the results of such studies by Herweijer, van den Engh and Nooter (1989) are shown in figure 2 where daunorubicin was used as the reporter probe for MDR in sensitive and resistant human ovarian cancer cell lines. These data have been redrawn from the original publication. The resistant line, A2780/R, was maintained continuously in medium containing daunorubicin at a concentration of 2 μM but the cells were incubated in drug-free medium for 48 hours prior to analysis. Time courses for the uptake of daunorubicin are shown for the sensitive (A2780/S) and resistant (A2780/R) lines and the former shows considerably greater uptake.

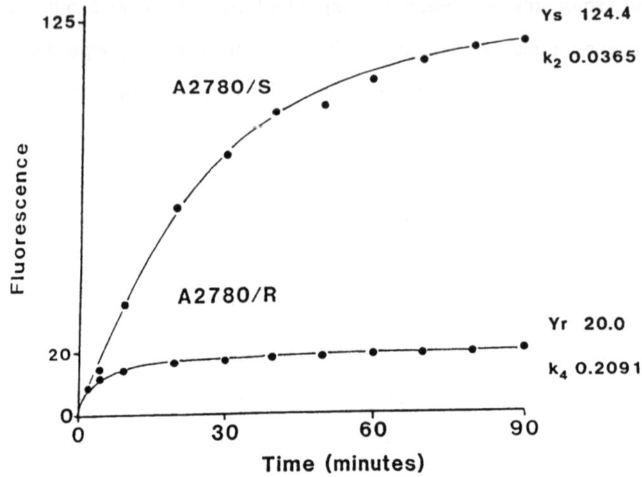

FIGURE 2) Daunorubicin time course uptake in sensitive (A2780/S) and resistant (A2780/R) cell lines. Redrawn from Herweijer, van den Engh and Nooter (1989).

These results are interesting not only for the effect they show but also for the information which can be obtained. It is clear that these uptake curves are of the inverted exponential type which suggests that the membrane traffic can be approximated with first order processes described by the following rate equation.

$$\frac{\delta S}{\delta t} = k_1 E - k_2 S$$

where S represents the intracellular drug concentration in the sensitive cells and E represents the extracellular concentration. The respective rate constants are k_1 for influx and k_2 for efflux. This equation can be rearranged to give,-

$$\frac{\delta S}{\delta t} + k_2 S = k_1 E$$

On integration this equation yields the familiar inverted exponential

$$S = \frac{k_1}{k_2} E (1 - e^{-k_2 t})$$

where k_2 is the rate constant for efflux and where the uptake curve should reach an asymptope, Y_0, of Exk_1/k_2. The comparable equation for resistant cells is

$$R = \frac{k_3}{k_4} E (1 - e^{-k_4 t})$$

where R is the intracellular drug concentration, k_3 is the rate constant for influx, k_4 is the rate constant for efflux and where the uptake curve should reach an asymptope of Exk_3/k_4.

The daunorubicin uptake results of Herweijer, van den Engh and Nooter (1989) shown in figure 2 were subjected to regression analysis using the inverted exponentials given above and the results are shown in table 1.

Cells	Y_0	Efflux rate constant	t	p
Sensitive	124.4	0.0365	0.32	> 0.60
Resistant	20.0	0.2091	2.12	> 0.05

TABLE 1
Asymptops and efflux rate constants for the sensitive and resistant cell lines with Student's t and p values.

The fit between the mathematical model and the data was good as can be seen from the values for Student's t and p in table 1 and in figure 2 where the curves are the inverted exponentials and the symbols are "our" data points measured from the figure in the authors' publication.

We can now see that the rate constant for efflux across the external membrane was about 6-fold greater in MDR resistant cells compared with the sensitive parent. However, what about the influx rate constants k_1 and k_3 ? We can see from the uptake equations that,-

$$Y_0 = \frac{k_1}{k_2} E \text{ for the sensitive cells and } Y_0 = \frac{k_3}{k_4} E \text{ for the resistant.}$$

Thus, $k_1\ E = Y_0 \times k_2 = 124.4 \times 0.0365 = 4.54$ (sensitive)
and, $k_3\ E = Y_0 \times k_4 = 20.0 \times 0.2091 = 4.18$ (resistant).

These two values, considering the potential inaccuracies of my measurements from the original figure particularly after enlargement on the photocopier, are remarkably close and within 8%. Thus, we can reasonably assume that the influx rate constant is the same in both cells and that the difference in fluorescence levels is due entirely to greater efflux in the resistant cells. It is also clear that if we can relate the fluorescence units of the Y_0 values to the extracellular drug concentration units, E, then we can also obtain the rate constant(s) for drug influx.

These conclusions were substantiated in the work of Herweijer, van den Engh and Nooter (1989) with the calcium channel blockers verapamil and cyclosporin A which inhibit gp^{170}. After addition of these agents to the resistant cells there was an immediate dose-dependent increase in daunorubicin accumulation which followed first order kinetics but, there was no comparable change in the sensitive cells.

3) Fluorescent reporter probes

The bisbenzimidazole DNA dye Hoechst-33342 can be used as a reporter probe for studying membrane transport and its uptake parallels that of a number of cytotoxic agents involved in multi-drug resistance and hence correlates with sensitivity to those agents (Lalande, Ling and Miller, 1981). Furthermore, Hoechst-33342 uptake is modified by verapamil (Krishan, 1987) and it is probable that the dye is secreted by cells via the gp^{170} pathway. The fluorescence emission spectrum from the DNA:Hoechst-33342 complex varies in different cell types and is dependent of the dye:DNA-phosphate ratio and pH (Watson et al, 1985; Smith, Nakeff and

Watson 1985). At low dye:DNA ratios Hoechst 33342 binds in the minor groove of DNA and is associated with a violet bias to the fluorescence emission. At higher dye:DNA ratios the dye stacks across the minor groove which is associated with quenching and hence a lower emission energy which results in a shift of the spectrum towards the red. Hence, cells which pump the dye out through the external cell membrane will exhibit low dye:DNA ratios and a violet bias to the emission.

Hoechst-33342 stained cells are UV excited and fluorescence is collected on three photomultipliers each guarded by 10 nm narrow band pass filters for simultaneous analysis in the violet (395nm-405nm), blue (495nm-505nm) and red (595nm-605nm) regions of the spectrum. The photomultiplier voltages, amplifier gains and thresholds of the instrument were set using control samples stained for 1 hour so that the easily defined G1 peak was placed at the same scale position on each photomultiplier analysis channel. Subsequent fluorescence from the test samples could then be compared with the control on each analysis channel to give a measure of relative changes in the emission spectrum. A total of 10^4 cells was analysed for each data set.

A number of multi-drug resistant cell lines have been developed in our laboratories and these were used in the investigation. Single cell suspensions were obtained by treating with 0.4% trypsin plus 0.02% versine and the concentrations were adjusted to $2x10^5$ cells ml^{-1} in growth medium. A stock solution of Hoechst-33342 was dissolved in distilled water at a concentration of 1mM. Aliquots of cells were stained by adding 10μl of the stock solution to 1 ml of the cell suspensions to give a final stain concentrations of 10μM. The cells were then incubated at 37^0C in the dark. Verapamil, when this was used, was added to give a final concentration of 3.3μM.

Time courses for the development of violet (400nm, left panel) and red (600nm, right panel) fluorescence in the H69/P and LX4 cell lines are shown in **figure 3** where the closed and open symbols represent the results from the sensitive (H69/P) and resistant (LX4) cell lines respectively.

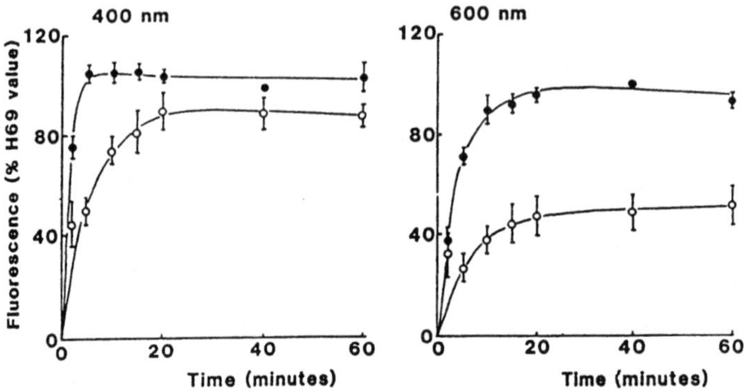

FIGURE 3) Time courses for the development of violet (400nm) and red (600nm) fluorescence in the H69/P and LX4 cell lines. The closed and open symbols represent the results from the parent (H69/P) and resistant (LX4) cell lines respectively. The error bars were drawn at the 95% confidence limits.

Note there is very much faster development of fluorescence in the parent and a considerable violet (400nm) bias in LX4 compared with its red (600nm) fluorescence. The instrument was setup with the parent cell line as the control and any differences in total fluorescence intensity arising from differences in cellular DNA content can be "normalized" by taking the red:violet ratio. This is shown in **figure 4** with the closed and open circles again representing the sensitive (H69/P) and resistant (LX4) cells. The lower ratio in the latter clearly demonstrates the considerable violet bias to the emission.

Figure 5 shows a time course for the development of fluorescence at three wavelengths (600nm, circles; 500nm, triangles; and 400nm squares) for H69/P and LX4 in left and right panels respectively where verapamil was added to both after 25 minutes exposure to Hoechst-33342. The solid symbols represent the results following verapamil treatment.

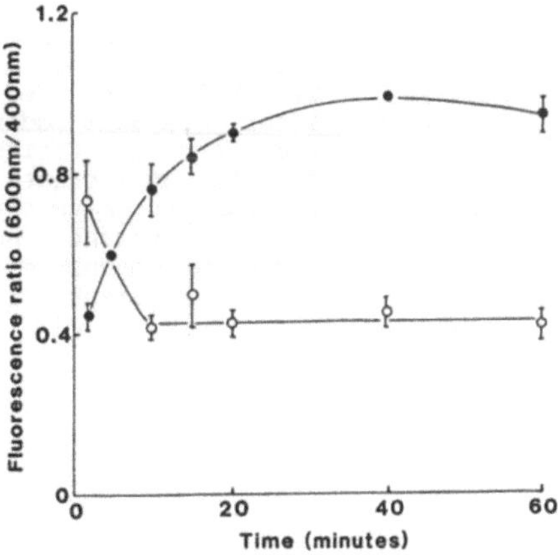

FIGURE 4) Red:violet ratios calculated from the data in **figure 3** where the closed and open symbols again represent the sensitive (H69/P) and resistant (LX4) cells.

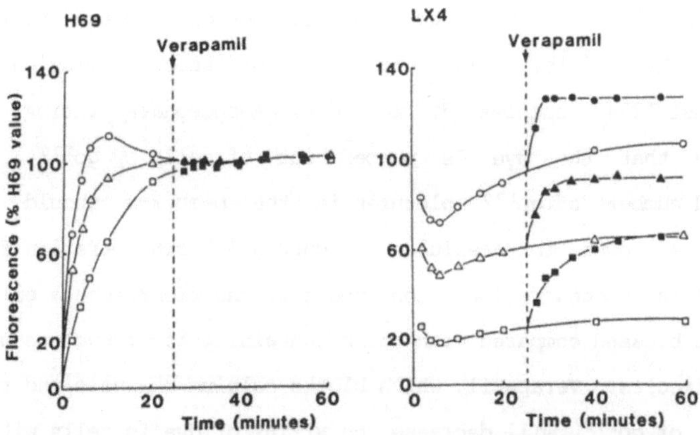

FIGURE 5) The time course for the development of fluorescence at three wavelengths (600nm, circles; 500nm, triangles; and 400nm squares) for H69/P and LX4 in the left and right panels respectively. Verapamil was added after 25 minutes exposure to Hoechst 33342. The solid symbols represent the results following verapamil exposure.

There was no difference in the sensitive cell line (left panel) but there was a pronounced change in the LX4 multi-drug resistant cell line (right panel) with considerable increase in fluorescence at all wavelengths. The red:violet ratios calculated from the data of **figure 5** are shown in **figure 6** where the red shift after verapamil addition in the LX4 cells (triangles) is very apparent however, there was no change in the sensitive parent (circles).

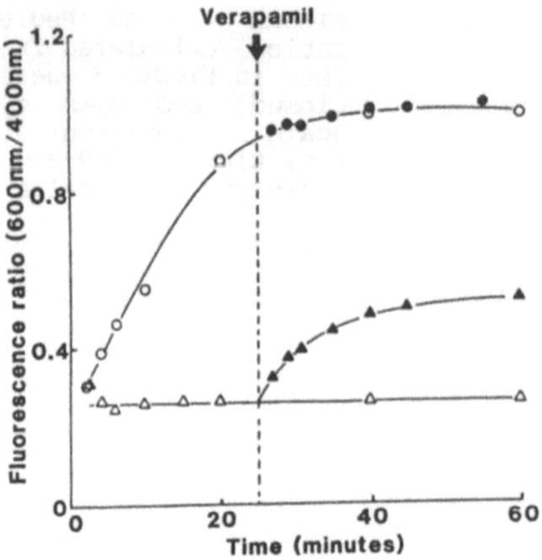

FIGURE 6) Red:violet ratios from the data in figure 5. The red shift after verapamil addition in the LX4 cells (triangles), due to an increase in the dye:DNA-phosphate ratio, is very apparent however, there was no change in the sensitive parent (circles).

This technique for assessing gp^{170} MDR depends on only two basic factors namely, the violet bias to the fluorescence emission from the DNA:Hoechst-33342 complex at low dye:DNA-phosphate ratios and the postulate that the dye is pumped out of cells by gp^{170}. Cells with increased numbers of gp^{170} molecules in the membrane should therefore, maintain a lower intracellular Hoechst-33342 concentration for a given extracellular concentration. Consequently, the fluorescence emission will be violet biassed compared with cells containing lower numbers of membrane gp^{170} molecules. Verapamil, which blocks calcium channels and reduces the activity of gp^{170}, will decrease the efflux of dye in cells with elevated gp^{170} levels. This results in a higher intracellular dye concentration, greater binding to DNA and a relative red shift to the emission spectrum. Cells with very low numbers of gp^{170} molecules will show little or no emission spectrum shift with verapamil.

4) Fluorescence quenching

The quenching of propidium iodide (PI) DNA fluorescence as a means of assessing nuclear adriamycin content was demonstrated by Krishan and Ganapathi (1979) but, this method is not particularly sensitive for dynamic studies. The major problem is that the range in uptake values and in enzyme kinetic determinations often cover many orders of magnitude and any assay involving loss of fluorescence due to quenching will rarely have a sufficiently large dynamic range. However, this may be the only way to study ligand binding if no other technique is possible. One such example is actinomycin-D which gives rise to quenching of fluorescence from DNA stained with acridine orange (Chambers and Watson, 1981).

5) Fluorescence analogues

Fluorescent analogues of the anti-folates, aminopterine and methotrexate, have been synthesised (Gapski et al, 1975; Henderson, Russell and Whiteley, 1980; Rosowsky et al, 1982; Kumar et al, 1983a,b) and uptake studies have been carried out with cells sensitive and resistant to methotrexate using one of its analogues. However, methotrexate resistance is different from that in MDR and two of the three mechanisms identified are attributable to Fisher (1961, 1962). The enzyme dihydrofolate reductase, DHFR, converts folic acid to folinic acid which is required in one of the pathways to DNA synthesis. Methotrexate binds to DHFR with an affinity about 40-fold greater than that of folic acid and blocks its action. Fisher (1961) demonstrated that methotrexate resistance could be caused by elevation of dihydrofolate reductase levels. Cells grown continuously in medium containing methotrexate exhibit amplification of the DHFR gene (Hanggi and Littlefield, 1976; Alt, Kellems and Schimke, 1976) where the multiple copies are contained in "double minute" chromosomes and are manifest in such cells at metaphase. Stable mutants are obtained when these double minutes become integrated into the normal chromosomes and the amplified gene can then be transmitted predictably from one generation to the next which confers considerable methotrexate resistance on the population. Fisher (1962) also demonstrated that the rate of influx of methotrexate (MTX) was about 15-fold lower in resistant compared with sensitive cells. At equilibrium the intracellular MTX concentration was between 10- and 30-fold lower in the resistant cells. However, this was not due to metabolic changes as dihydrofolate reductase levels were equal in both the resistant and sensitive cells. Changes in methotrexate membrane transport contributing to resistance (Schimke, 1984) have also been demonstrated by Sirontak et al (1981) and Dembo, Sirotnak and Moccia (1984) have presented evidence that the cellular influx and efflux pathways for methotrexate are different. The third mechanism of methotrexate resistsnce is due to an alteration of the affinity of methotrexate binding to DHFR (Flintoff, Davidson and Siminovitch, 1976; Haber et al, 1981).

To date relatively limited work has been carried out on methotrexate resistant cells using flow cytometry. This has been due in part to the difficulty of producing fluorescent analogues with low background as pointed out by Gaudray, Trotter and Wahl (1986). However, these authors overcame this probem with a fluorescenated derivative of the ligand and presented data which suggested that staining heterogenity with the analogue was partly due to transport variability within the population. Assaraf and Schimke (1987) and Assaraf, Seamer and Schimke (1989) loaded cells with fluorescent methotrexate then examined the ability of hydrophilic and lipophilic antifolates to displace the probe. Methotrexate resistant cells incubated with methotrexate (hydrophilc) exhibited no reduction in fluorescence compared with sensitive cells however, incubation with lipophilic analogues produced reductions in fluorescence in both sensitive and resistant cells. This is very compelling evidence for a membrane active-transport mechanism for methotrexate and hydrophilic analogues which is deficient in resistant cells. The lipophilic analogues enter cells by passive diffusion without the need for an active-transport mechanism hence, these displaced the fluorescenated probe bound to DHFR in both the sensitive and resistant cell lines. Similar types of studies have been carried out by Rosowsky et al (1986) and Wright et al (1987). **Figure 7** shows the uptake of fluorescent methotrexate versus time in a parental chinese hamster cell line, AA8, and in a DHFR-deficient cell line, DG-44.

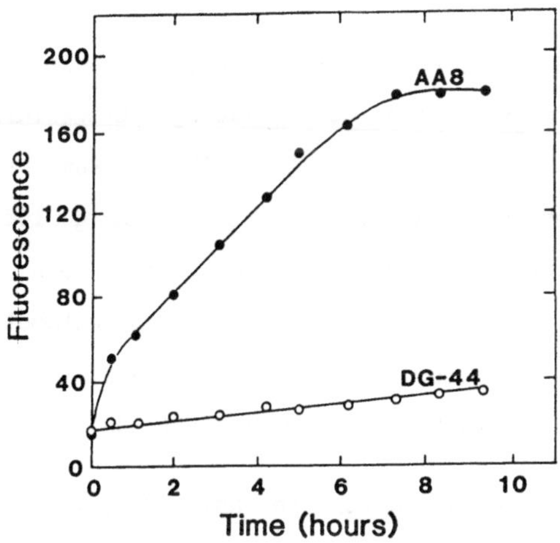

FIGURE 7) Uptake kinetics for fluorescent methotrexate in a parental cell line (AA8, O) and a dihydrofolate reductase deficient cell line (DG-44, O) from Assaraf, Seamer and Schimke (1989).

The former exhibits very considerably greater fluorescence which is attained at a more rapid rate. The effect of a temperature reduction, which slows the uptake rate of active transport processes but not of passive diffusion, is shown in **figure 8**. There was a profound decrease in fluorescent methotrexate uptake in parental CHO cells when the temperature was decreased from 37°C to 4°C, an effect which was easily observed in the initial stages of uptake with a temperature decrease to 27°C. Both of these last two data sets were redrawn from Assaraf, Seamer and Schimke (1989).

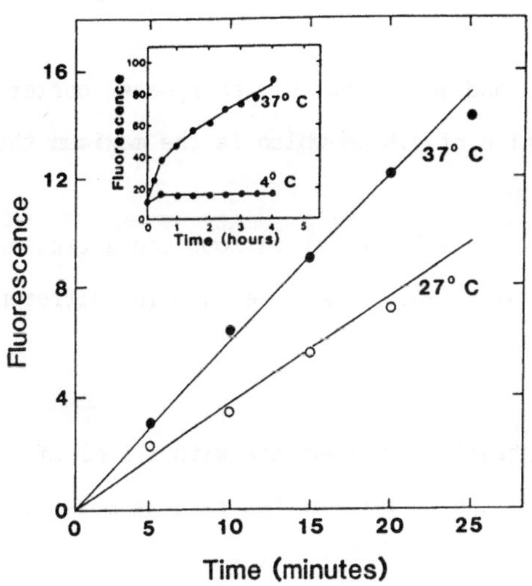

FIGURE 8) Temperature dependent uptake of fluorescent methotrexate in parental CHO cells. The insert shows that cooling from 37°C to 4°C effectively abolishes uptake. The main graph shows that a reduction in uptake can be measured with a temperature decrease to only 27°C, from Assaraf, Seamer and Schimke (1989).

6) <u>Enzyme inhibition kinetics</u>

The chloroethylnitrosoureas are the only agents where enzyme inhibition kinetics have been used as an indirect method of estimating intracellular accumulation. The nitrosoureas, CCNU, BCNU (Begleiter, Lam and Goldenberg, 1977) and chlorozotocin (Lam, Talgoy and Goldenberg, 1980) gain access to the cell interior by passive diffusion as the uptake of each is unsaturable, independent of temperature and unaltered by metabolic inhibitors. Esterase activity is inhibited by nitrosoureas and related isocyanates due to carbamoylation (Dive, Workman and Watson (1988).

Esterase activity, and hence inhibition, can be assayed using the lipophilic non-fluorescent substrate fluorescein diacetate, FDA (Rotman and Papermaster, 1966), which releases the fluorescent product fluorescein due to esterase activity. As FDA is lipophillic and it must first be dissolved in acetone before addition to the physiological reaction mixture and a protocol is given below.

1) Prepare the single cell suspension in heat inactivated medium and adjust to 10^6 cells ml $^{-1}$. Heat inactivated medium is needed as serum, particularly foetal calf, can contain considerable concentrations of esterases.

2) Dissolve pure AR grade FDA in 1.0 ml spectrograde acetone to give a stock solution at a concentration of 12 mM. This should be kept in the dark at -10^0C.

3) Take 20 μl of the stock solution and add to 50 ml phosphate buffered saline. This gives a concentration of 4.8 μM which is the maximum that can be obtained without flocculation.

4) 1.5 ml aliquots of working substrate solution at various concentrations are now prepared by dilution with PBS. We use 6 - 10 different concentrations.

5) Tune the laser to 488 nm.

6) Mix 1.5 ml of the highest concentration of substrate with 1.5 ml of the cell suspension and introduce into the instrument "as fast as possible".

Set up the instrument so that about half the population is still "on-scale" after the longest time interval to be investigated.

7) Flush the tubing with dilute bleach, FDA is adsorbed to plastic, then flush through with water to remove the bleach.

8) Start the assay with the lowest substrate concentration.

9) The kinetic measurements you can obtain will depend on the capability of your flow cytometer. With primitive instruments, by which we mean all those which do not incorporate the time stamp from the computer clock for each cell into the data base, you will have to record a green fluorescence histogram of say 10000 cells at known intervals.

10) Repeat with increasing substrate concentrations.

11) Plot the medians of the fluorescence distributions against the time after mixing cells with substrate to generate a progress curves for each substrate concentration.

12) In instruments which incorporate the computer time stamp in the data base the progress curve for each substrate concentration is obtained by plotting the fluorescence from each cell directly against time.

13) Obtain the initial reaction velocity by regression analysis of the medians of the fluorescence distributions on time for the discrete points obtained at step 11 or by regression analysis of fluorescence from each cell on time obtained at step 12.

Esterase inhibition can now be assayed in a similar manner after exposure of cells to the inhibiting agent. **Figure 9** shows three panels where the top two diagrams (panel A) show size versus fluorescence on the X and Z horizontal axes with time (which increases upwards on these displays) on the Y-axis. Contour plots of size versus fluorescence for 10 equal "time-slices" are shown within these 3-dimensional data spaces. The top left is the control and top right shows the inhibition due to BCNU. Panel B shows the medians of the fluorescence distributions of each "time-slice" plotted against time.

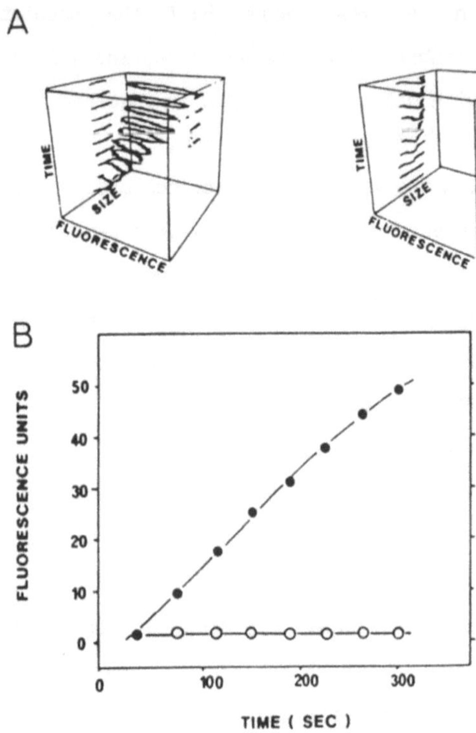

FIGURE 9) Inhibition of esterase activity by BCNU-derived isocyanate (Dive, Workman and Watson, 1987b). The top two diagrams (panel A) show size versus fluorescence on the X and Z horizontal axes with time (which increases upwards on these displays) on the Y-axis. Contour plots of size versus fluorescence for 10 equal "time-slices" are shown within these 3-dimensional data spaces. Top left is the control and top right shows the inhibition due to BCNU. Panel B shows the medians of the fluorescence distributions of each "time-slice" plotted against time.

By carrying out the determinations with multiple inhibitor concentration it is possible to estimate the concentration which will reduce esterase activity to 50% of that of the control, the I_{50} value, as shown in **figure 10**. Comparisons between agents can then be made using the I_{50} value. Similar assays were performed with a number of drugs using both flow cytometry for intact cells and conventional spectrofluorimetry with sonicated preparations and the I_{50} values for the two methods are compared in table 2 together with the ratio of results from intact cells to those in sonicates.

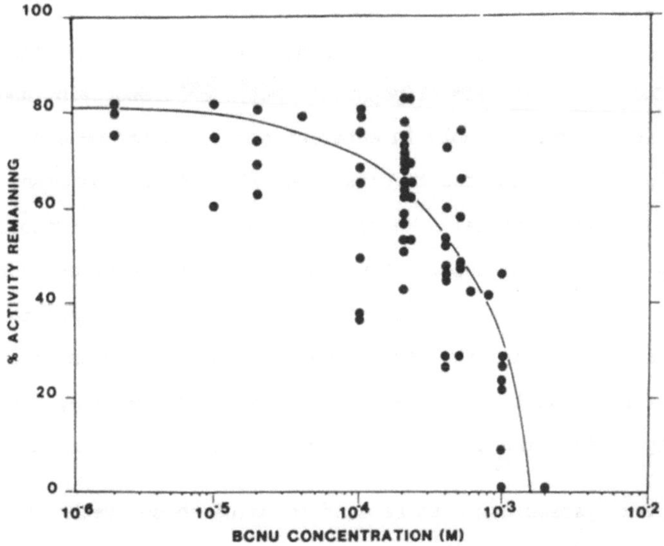

FIGURE 10) Enzyme activivity for multiple concentrations BCNU. The dose which produces 50% inhibition of enzyme activity, I_{50}, can be estimated from data such as these.

I_{50} values, (M)

Drug	Intact cells	Sonicates	Ratio
CHI	1.1×10^{-5}	2.6×10^{-5}	0.423
CCNU	3.8×10^{-5}	8.3×10^{-5}	0.458
BCNU	5.0×10^{-5}	9.2×10^{-5}	0.543
CEI	1.0×10^{-5}	1.6×10^{-5}	0.625
TCNU	1.8×10^{-4}	9.9×10^{-5}	1.818
ACNU	7.3×10^{-3}	3.2×10^{-3}	2.281
CHLOZ	$>1.5 \times 10^{-2}$	2.4×10^{-3}	>6.250
GANU	$>1.5 \times 10^{-2}$	2.4×10^{-3}	>6.250

TABLE 2
I_{50} values for a variety of nitrosoureas and related isocyanates.

The results in table 2 are interesting. The I_{50} ratios for CHI, CCNU, BCNU and CEI indicate that only about half the concentration of drug is required to obtain 50% esterase activity inhibition in whole cells compared with sonicates. With TCNU and ACNU this is reversed with double the drug concentration required in whole cells compared with sonicates however, with chlorozotozin and GANU this has increased to >6-fold. These results almost exactly parallel the hydrophilicity of the agents. CHI, CCNU, BCNU and CEI are lipophilic but, chlorozotozin and GANU are hydrophilic, readily soluble in water at physiological pH and temperature and would not be expected to cross the external cell membrane without an active transport mechanism. We have seen earlier that nitrosoureas enter cells by passive diffusion thus, we would not expect chlorozotozin and GANU to enter cells. This conclusion would seem to be substantiated by the data in table 2 and the method seems capable of assessing not only the ability of such agents to cross the external cell membrane, but also their capability of inducing damage having entered the cell.

Complex systems can also be studied using these techniques. **Figure** 11 shows forward versus 90^0 light scatter from mouse bone marrow where the major cluster is surrounded by an oval gate.

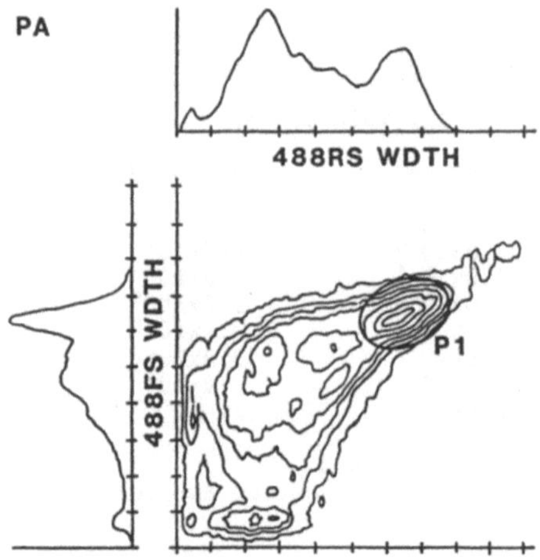

FIGURE 11) Contour plots of forward scatter on the ordinate versus 90^0 scatter on the abscissa from mouse bone marrow with the associated mono-dimensional histograms adjacent to the respective axes. The major population is gated and marked as P1.

The esterase activity associated with this gate is shown in figure 12 where three subsets, labelled S1, S2 and S3, can be identified based on their esterase activity indicated by increasing green fluorescence (520RF AREA) with time.

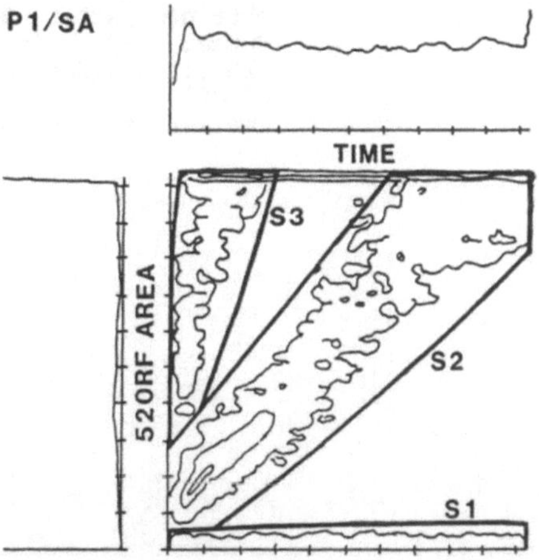

FIGURE 12) Esterase activity where green fluorescence is plotted versus time for the gated population of figure 11 and three subsets, with high, medium and low activity are apparent.

The marrow was then treated with BCNU at a concentration of 10^{-4} M and the results are shown in figure 13. Subsets S2 (medium activity) and S3 (highest activity) are depicted by the dotted and uninterrupted lines respectivley. The circles represent the untreated control cells and the squares represent the data after BCNU treatment. Both subsets exhibited a decrease in esterase activity following BCNU treatment but the reductions in activity were different in the two subsets. The slopes were compared after linear regression analysis and subset S2 (dotted lines) exhibited a reduction of 80% whilst subset S3 exhibited a smaller reduction of 55%. This type of effect, with different subsets behaving differently, was a consistent finding for almost all the populations definable in the forward versus 90^0 scatter data space. This is also compatible with the work of Sonneveld and van den Engh (1981) where myeloid leukaemia progenitor cells exhibited different uptake and sensitivity to different anthracyclines.

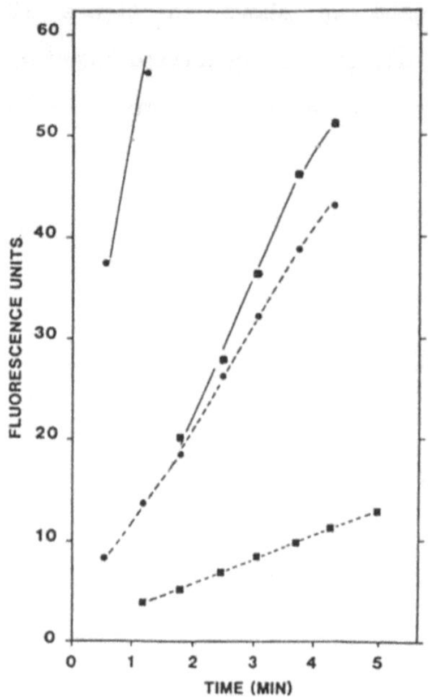

FIGURE 13) Esterase activity assayed after treatment of mouse bone marrow with 10^{-4}M BCNU for 60 minutes. The dotted and solid lines respectively represent subsets S2 and S3 of figure 11. The circles represent the data from control cells and the squares were obtained after BCNU treatment. Subset S2 (dotted lines) exhibited a decrease in activity of 80% whilst S3 exhibited a smaller redution of 55%.

REFERENCES

Alt, F.W., Kellems, R.E. and Schimke, R.T. (1976) "Synthesis and degradation of folate reductase in sensitive and methotrexate resistant lines of S-180 cells"
J. BIOL CHEM. 251; 3063-3074

Assaraf, Y.G., Seamer, L.C. and Schimke, R.T. (1989) "Characterization by flow cytometry of fluorescein-methotrexate transport in chinese hampster ovary cells"
CYTOMETRY 10; 50-55

Assaraf, Y.G. and Schimke, R.T. (1987) "Identification of methotrexate transport deficiency in mammalian cells using fluoresceinated methotrexate and flow cytometry"
PROC. NATNL. ACAD. SCI. USA. 84; 7154-7158

Begleiter, A., Lam, H.Y.P. and Goldenberg, G.J. (1977) "Mechanism of uptake of nitrosoureas by L5178Y lymphoblasts in vitro"
CANCER RES. 37; 1022-1027

Chambers, S.H. and Watson, J.V. (1981) "Effect of actinomycin-D on the staining of whole cells with acridine orange"
J. HISTOCHEM. CYTOCHEM. 29 644-648

Dembo, M., Sirotnak, F.M. and Moccia, D.M. (1984) "Effects of metabolic depravation on methotrexate transport in L1210 leukaemia cells: further evidence for separate influx and efflux systems and different energetic requirements"
J. MEMBRANE BIOL. 78; 9-17

Dive, C., Workman, P. and Watson, J.V. (1988) "Inhibition of intracellular esterases by antitumour chloroethylnitrosoureas: Measurement by flow cytometry and correlation with molecular carbamoylation activity"
BIOCHEM. PHARMACOL. 37; 3987-3993

Durand, R. and Olive, P. (1981) "Flow cytometric studies of intracellular adriamycin in single cells in vitro"
CANCER RES. 41; 3489-3494

Fisher, G.A. (1961) "Increased levels of folic acid reductase as a mechanism of resistance to amethopterin in leukaemic cells"
BIOCHEM. PHARMACOL. 7; 75-80

Fisher, G.A. (1962) "Defective transport of amethopterin (methotrexate) as a mechanism of resistance to the antimetabolite in L5178Y leukaemia cells"
BIOCHEM. PHARMACOL. 11; 1233-1234

Flintoff, W.E., Davidson, S.V. and Siminovitch, L. (1976) "Isolation and partial characterization of three methotrexate-resistant phenotypes from chinese hamster ovary cells"
SOMATIC CELL GENET. 2; 245-261

Gapski, G.R., Whiteley, J.M., Rader, J.I., Cramer, P.L. Henderson, G.B., Neef, V. and Huennekens, F.M. (1975) "Synthesis of a fluorescent derivative of amethopterin"
J. MED. CHEM. 18; 526-528

Gaudray, P., Trotter, J. and Wahl, G.M. (1986) "Fluorescent methotrexate labelling and flow cytometric analysis of cells containing low levels of dihydrofolate reductase"
J. BIOL. CHEM. 261; 6285-6292

Haber, D.A., Beverly, S.M., Kiely, M.L. and Schimke, R.T. (1981) "Properties of an altered dihydrofolate reducatase encoded by amplified genes in cultured mouse fibroblasts"
J. CELL BIOL. 256; 9501-9510

Hanggi, U.J. and Littlefield, J.W. (1976) "Altered gene regulation of the rate of synthesis of dihydrofolate reductase in methotraxate-resistant hamster cells"
J. BIOL. CHEM. 251; 3075-3082

Henderson, G.B., Russell, A. and Whiteley, J.M. (1980) "A fluorescent derivative of methotrexate as an intracellular marker of dihyrdofolate reducatase in L1210 cells"
ARCH. BIOCHEM. BIOPHYS. 202; 29-34

Herweijer, H., van den Engh, C.J. and Nooter, K. (1989) "A rapid and sensitive flow cytometric method for the detection of multidrug-resistant cells"
CYTOMETRY 10; 463-468

Keij, J.F., Griffioen, A.W., The, T.H. and Rijkers, G.T. (1989) "INCA: software for Consort 30 analysis of flow cytometric calcium determinations"
CYTOMETRY 10; 814-817

Krishan, A. (1987) "Effects of drug efflux blockers on vital staining of cellular DNA with Hoechst 33342"
CYTOMETRY 8; 642-645

Krishan, A. and Ganapathi, R. (1979) "Laser flow cytometry and cancer chemotherapy: detection of intracellular anthracyclines by flow cytometry"
J. HISTOCHEM. CYTOCHEM. 27; 1655-1656

Krishan, A. and Ganapathi, R. (1980) "Laser flow cytometric studies on intracellular fluorescence of anthracyclines"
CANCER RES. 40; 3895-3900

Krishan, A., Sauerteig, A., Gordon, K and Swinkin, C. (1986) "Flow cytometric monitoring of cellular anthracycline accumulation in murine leukaemia cells"
CANCER RES. 46; 1768-1773

Krishan, A., Sridhar, K.S., Davila, E., Vogel, C. and Sternheim, W. (1987) "Patterns of anthracycline retention modulation in human tumour cells"
CYTOMETRY 8; 306-314

Kumar, A.A., Freisheim, J.H., Kempton, R.J., Anstead, G.M. Black, A.M. and Judge, L. (1983a) "Synthesis and characterization of a fluorescent analogue of methotrexate"
J. MED. CHEM. 26; 111-113

Kumar, A.A., Kempton, R.J., Anstead, G.M., Proce, E.M. and Freisheim, J.H. (1983b) "High-performance liquid chromatography of methotrexate analogues containing terminal lysine or ornithine and their dansyl derivatives"
ANAL. BIOCHEM. 128; 191-195

Lalande, M., Ling, V. and Miller, R.G. (1981) "Hoechst 33342 dye uptake as a probe of membrane permeability changes in mammalian cells"
PROC. NATNL. ACAD. SCI. USA. 78; 363-367

Lam, H.Y.P., Talgoy, M.M. and Goldenberg, G.J. (1980) "Uptake and decomposition of chlorozotocin in L5178Y lymphoblasts in vitro"
CANCER RES. 40; 3950-3955

Martin, J.C. and Swartzendruber, D.E. (1980) "Time: a new parameter for kinetic measurement in flow cytometry"
SCIENCE 207; 199-200

Morgan, S.A., Watson, J.V., Twentyman, P.R. and Smith, P.J. (1989) "Flow cytometric analysis of Hoechst 33342 uptake as an indicator of multi-drug resistance in human lung cancer"
BRIT. J. CANCER 60; 282-287

Nooter, K., Oostrum, R., Jonker, R.R., van Dekken, H., Stokdijk, W. and van den Engh, C.J. (1989) "Effects of cyclosporin A on daunorubicin accumulation in multidrug-resistant P388 leukaemia cells measured by real-time flow cytometry"
CANCER CHEMOTHER. PHARMACOL. 23; 296-300

Nooter, K., van den Engh, C.J. and Sonneveld, P. (1983) "Quantitative flow cytometric determination of anthracycline content of rat bone marrow cells"
CANCER RES. 43; 5126-5130

Rosowsky, A., Wright, J.E., Cucchi, C.A., Boeheim, K. and Frei, E. (1986) "Transport of a fluorescent antifolate by methotrexate-sensitive and methotrexate-resistant human leukaemic lymphoblasts"
BIOCHEM. PHARMACOL. 35; 356-360

Rosowsky, A., Wright, J.E., Shapiro, H., Beardsley, P. and Lazarus, H. (1982) "A new fluorescent dihyrdofolate reductase probe for studies of methotrexate resistance"
J. BIOL. CHEM. 257; 14162-14167

Ross, D.D., Joneckis, C.C. and Schiffer, C.A. (1986) "Effects of verapamil on the in vitro intracellular accumulation and retention of daunorubicin in blast cells from patients with acute non lymphatic leukaemia"
BLOOD 68; 83-88

Ross, D.D., Ordonez, J.V., Joneckis, C.C., Testa, J.R. and Thompson, B.W. (1988) "Isolation of highly multidrug resistant P388 cells from drug sensitive P388/S cells by flow cytometric cell sorting"
CYTOMETRY 9; 359-367

Rotman, B. and Papermaster, B.W. (1966) "Membrane properties of living mammalian cells as studied by enzymatic hydrolysis of fluorogenic esters"
PROC. NATNL. ACAD. SCI. USA. 55; 134-141

Schimke, R.T. (1984) "Gene amplification, drug resistance and cancer"
CANCER RES. 44; 1735-1742

Sirontak, F.M., Moccio, D.M., Kelleher, L.E., and Goutas, L.J. (1981) "Relative frequency and kinetic properties of transport-defective phenotypes among methotrexate-resistant L1210 clonal cell lines derived in vitro"
CANCER RES. 41; 4447-4452

Smith, P.J., Nakeff, A. and Watson, J.V., (1985) "Flow-cytometric detection of changes in the fluorescence emission spectrum of a vital DNA-specific dye in human tumour cells"
EXPTL. CELL RES. 159; 37-46

Sonneveld, P. and van den Engh, C.J. (1981) "Differences in uptake of adriamycin and daunomycin by normal BM and acute leukaemia cells determined by flow cytometry"
LEUKAEMIA RES. 5; 251-257

Tsuro, T., Iida, H., Tsuragoshi, S. and Sakurai, Y. (1982) "Increased accumulation of vincristine and adriamycin in drug-resistant P388 tumour cells following incubation with calcium antagonists and calmodulin inhibitors"
CANCER RES. 42; 4730-4733

Visser, J.W.M., van den Engh, G.J. and van Bekkum, D.M. (1980) "Light scattering properties of murine haemopoetic stem cells"
BLOOD CELLS 6; 391-407

Watson, J.V., Chambers, S.H., Workman, P. and Horsnell, T.S. (1977) "A flow cytometric method for measuring enzyme reaction kinetics in populations of single cells"
FEBS LETT. 81; 179-192

Watson, J.V., Morgan, S.A. and Dive, C.D. (1989) "Drug resistance analysis in populations of individual cells using flow cytometry"
Chapter 14 in "Ovarian Cancer; biological and therapeutic challenges" Eds Sharp, F., Mason, W.P. and Leake, R.E., CHAPMAN and HALL MEDICAL, LONDON

Watson, J.V., Morgan, S.A. and Smith, P.J. (1989) "Analysis of the Hoechst-33342 fluorescence emission spectrum in individual cells with reference to multi-drug resistance"
PROTIDES of the BIOLOGICAL FLUIDS 36; 359-372

Watson, J.V., Nakeff, A., Chambers, S.H. and Smith, P.J. (1985) "Fluorescence emission spectrum analysis of chicken thymocytes using Hoechst-33342"
CYTOMETRY 6; 310-315

Wright, J.E., Rosowsky, A., Boeheim, K., Cucchi, C.A. and Frei, E. (1987) "Flow cytometric studies of methotrexate resistance in human squamous carcinoma cell cultures"
BIOCHEM. PHARMACOL. 36; 1561-1564

Endogenous lectins in circulating cells and their glycosylated ligands : their role in T cell homing

Véronique Denis, Michèle Mitterand and Claudine Kieda

CNRS - Centre de Biophysique Moléculaire - Glycobiologie

1, rue Haute, 45071 Orléans, France

The past few years have permitted a remarkable expansion of our knowledge on cell adhesion and recognition which results mainly from three steps in the progress of scientific acquisitions : the burst of interest about adhesion molecules, the knowledge of the key part played by endothelium and endothelial cells activity as well as, the role of lectins and their glycosylated ligands in the establishment of specific intercellular recognition-dependent events in both normal biological responses and pathological processes leading to homing of normal cells among lymphoid sites during the immune responses or to damageous localization of tumor cells among tissues where they metastasize.
Numerous adhesion molecules have been described. They had to be classified into three main famillies (see the review by Springer, 1990). The several steps which co-operate to produce an efficient adhesion were attempted to be deciphered (Butcher, 1991). This established the role of lectin (selectin) molecules in binding their glycoconjugate ligands as being the first event of the adhesion cascade which allows braking, rolling and arrest of blood borne circulating cells onto the endothelial monolayer (Picker, 1992), to finally enter the underlying tissue.
Complex glycannic structures, and their substitutions by phosphoryl or sulphate groups are lectins ligands in the leukocyte-endothelial cell recognitions (selectins) or in infectious diseases (viral lectins).

Lectins as homing signal molecules :

The study of lymphocyte homing has revealed the importance of lectin-oside recognitions in cellular interaction (Butcher *et al.,* 1980) and the selectin familly (Bevilacqua *et al.* 1989, Johnston, *et al.* 1989, Siegelman *et al.* 1989, Lasky *et al.* 1989) : * the E-selectin on cytokine-activated endothelial cells, **the murine homing receptor, designated L-selectin on leukocytes; these two molecules are known to participate to leukocyte-endothelial adhesive events in various settings while, *** the third one, known as platelet activation-dependent granule protein, P-selectin is also present in endothelial cells after activation by thrombin (Mc Ever *et al.,* 1989). In parallel, biochemical studies on lymphocyte maturation had discovered and demonstrated the existence of lectin-like molecules (Kieda *et al.,* 1978, 1979).

In vivo homing could be mainly studied by means of **flow cytometry**. The direct role of lectins in lymphocyte homing could be demonstrated by *in vivo* experiments tracking the behaviour of *lectin distinct* cell subpopulations. Cells had been *ex vivo* separated on the basis of the type of lectin they express on their plasma membrane and further reinjected. While conducting such experiments, it also directly showed

NATO ASI Series, Vol. H 95
Flow and Image Cytometry
Edited by Alain Jacquemin-Sablon

the participation of endothelial cell membrane lectins in the homing pattern of lymphocytes (Kieda *et al.,* 1993).

Bargatze *et al.*(1987), showed that lymphoma cells express receptors specific for endothelial cells of post-capillary venules. Endothelial cells properties are a function of their localization. Furthermore, their surface antigenic molecules and properties are highly modulable depending upon the cytokine environment of endothelial cells. This work contributed to establish the role of endothelium in arresting blood borne cells. Furthermore, it indicated that tumor cells metastases might be established as a result of the same mechanisms which allow normal cells to achieve proper homing. This was in accordance with our hypothesis of a *double lectin- glycoconjugate recognition event* to set endothelial cell/circulating cell specific adhesion. This has been prompted by data obtained with tumor cells and host tissue endothelial cells (Kieda and Monsigny, 1986).

The **double lectin-glycoconjugate recognition hypothesis** is being demonstrated in various instances and the easiest way to attempt a molecular approach is an *in vitro* cellular model for circulating cell/endothelial cell adhesion and further extravasation. Such experiments require endothelial cells which reflect the *in vivo* reality; cell lines have been developed in this purpose (Bizouarne *et al.,* 1993 a,b). This has allowed to demonstrate that endothelial cells from different lymphoid organs express different membrane lectins and to show further that lectins as adhesion molecules are expressed on their surface as a function of the cytokine microenvironment. From these properties depends the resulting specificity of homing. Such biological implications of glycans open new therapeutics based on specific osidic sequences.

***In vitro* cellular model for the study of cell adhesion mechanism in homing : Flow cytometry assessment and identification.**

☞ Lymphocytes actively adhere to endothelial cells. An in vitro model for lymphocyte homing was designed using frosen sections of lymphoid organs and following the adhesion of lymphocytes. This showed the selectivity of adhesion towards endothelial cells.

☞ Endothelial cells differ according to the tissue they originate from. They express different receptors on their surface, namely do they express different addressins i.e. MECA 79 on peripheral lymph node endolhelial cells and MECA 367 on Peyer's patches endothelial cells; they also express different sugar specific receptors i. e. fucose and 6-phospho mannose- specific receptors, respectively.

☞ A selection assay based on the selective adhesion of cells onto monolayers of endothelial cells has been designed to approach the molecular mechanisms of invasive normal and/or transformed cells.

It consists of :

- labeling and culture of endothelial cells;
- labelling cells to adhere ;
- adhesion step: quantitative and qualitative ;
- recovery of the adherent and non adherent cells ;
- quantification by fluorescence based plate reader;
- identification of the cell populations.

Fluorescent labelling of cells, Adhesion, Identification.

1. HECa10 endothelial cells are labeled using PKH as a fluorescent dye :
Cells were seeded 4 hours before the assay in order to work with exponentially growing cells. They were previously labelled by PKH 26GLaccording to Horan and Slezak, 1990, Nature, 340, 167-168 .
2 X 107 cells were resuspended in one milliliter of RPMI medium without serum and PKH 26GL is diluted 1/250 in diluent C at 37oC. Labelling is achieved by mixing 1: 1 (v/v) cell supension to dye solution and incubating for 3 minutes at 25oC under gentle stirring. Labeling is stopped by adding an equal volume of foetal bovine serum, incubation is persued for 5 min at 25oC. Then the cells are washed by medium and centrifugation.
After washing, cells are seeded in Limbro 24 wells culture plates. 4 X 104 cells per well cultured for 24 hours before the assay.

2. Lymphoid cells labelling by Calcein :
The cell labeling is performed according to the method described by Akeson and Woods (J. immunol. Meth., 1993, 163:181-185). Calcein AM (Molecular Probes, Eugene, OR) stock solution in DMSO 1mg/ml. It is diluted immediately before use in 500 μl of medium without serum at 37oC for 10 min. to give a final concentration 200μM (2.0 % DMSO).
To the cell suspension, adjusted at 5 X 106 cell/ml, calcein AM (from the stock solution is added ata final concentration 7,5 μM at 37oC for 30 min.
Labelling is stopped by addition of medium containing serum.

3. Adhesion step :
Adhesion is performed in a routine assay by adding various proportions of lymphoid cells to endothelial cells (from 1/1 to 10/1). Adhesion is allowed for various time lengths at 4oC and/or at 37oC. After which non adherent cells are recovered separately. Both adherent and non adherent fractions are lysed by adding 100μl of 0.1% Triton X IOO in PBS. Fractions are transferred fo fluoroscan (Labsystem) plates and read with properly adjusted filter setting for PKH (λex = 544 nm, λem = 590 nm) and Calcein AM (λex = 485nm, λem = 530 nm). Quantification is done by estimating the cell concentration relative to standard curves.

4. Identification of the cell populations adhering to endothelial cells. In a whole population of lymphoid cells a subpopulation is selectively adhering to one type of

endothelial and this is dependenl upon the biologioca state and the cytokine environment. It is possible to identify the cells which are edhering to endothelial cells.
After the adhesion step:
- recover non adhering cells ,
- remove adhering cells by treatment with EDTA ;
- label both populations by monoclonal antibodies anti CD and/or neoglycoproteins ;
- FACS analysis of both populations allow the identification.

Endothelial cells are easily identified by their scatters and their labelling by PKH. Lymphoid cells are not previously labelled in this case and then idcntified by their classical surface marker, such as for T cells : CD3, CD4, CD8 ; or their sugar specific receptors (membrane lectins) labelled by **fluorescent neoglycoproteins** (Monsigny *et al.* 1984, Biol. Cell, 51, 187-196).

References:

Bargatze, R.F.,Wu, N.W., Weissman, I.L. and Butcher, E.C. 1987, High endothelial veinules binding as a predictor of the dissemination of passaged murine lymphomas. J. *Exp. Med.,* 166: 1125-1131.

Bevilacqua, M.P. Stengelin, S. Gimbrone, M. A. Jr and Seed, B. 1989, Endothelial adhesion molecule 1 : an inducible receptor for neutrophils related to complement regulatory proteins and lectins. *Science,* 243:1160-1165.

Bizouarne, N., Denis, V., Legrand, A., Monsigny, M. and Kieda, C. 1993, A SV40 immortalized murine endothelial cell line from peripheral lymph node high endothelium expresses a newα-L-fucose binding protein. *Biol. Cell,* 79: 209-218.

Bizouarne, N., Mitterrand, M., Monsigny, M. and Kieda, C. 1993, Characterization of membrane sugar-specific receptors in cultured high endothelial cells from mouse peripheral lymph nodes. *Biol. Cell,* 79: 27-37.

Butcher, E. C. 1991, Leukocyte-endothelial cell recognition: three (or more) steps to specificity and diversity. *Cell,* 67: 1033-1037.

Butcher, E.C., Scollay, R.G. and Weissmann, I.L., 1980, Organ specificity of lymphocyte migration: mediation by highly selective lymphocyte interaction with organspecific determinants on high endothelial venules. *Eur. J. Immuno1.10:556-561.*

Johnston, G.I., Cook, R. G., Mc Ever, R.P. 1989, Cloning of GMP-140, a granule membrane protein of platelets and endothelium: sequence similirarity to proteins involved in cell adhesion and inflammation. *Cell,* 56: 1033-1044 .

Kieda, C., Bizouarne, N., Denis, V. and Mitterrand, M., 1993 T lymphocytes recognition molecules in homing: a flow cytometry study of lectin-glycoconjugates interactions. *NATO ASI* series, H67: 155-164

Kieda, C., Bowles, D. J., Ravid, A., Sharon, N., 1978, Lectins in lymphocyte membrane. *FEBS Letters, 94:* 391-396.

Kieda, C. and Monsigny, M 1986, Involvement of sugar receptors and membrane glycoconjugates in the adhesion of 3LL subpopulations to cultured pulmonary cells. *Invasion Metastasis,* 6:347-366.

Kieda, C., Roche, A. C., Delmotte, F., Monsigny, M. 1979, Lymphocyte membrane lectins. Direct visualization by the use of fluoresceinyl glycosylated cytochemical markers. *FEBS Letters,* 99:329-332.

Lasky, L.A., Singer, M. S., Yednock, T. A., Dowbenko, D., Fennie, C., Rodriguez, H. Nguyen, T, Stachel, S. and Rosen, S.D., 1989, Cloning of a lymphocyte homing receptor reveals a lectin domain. Cell 56: 1045-1051.

Mc Ever, R. P., Beckstead, J.H., Moore, K. L., Marshall-Carlson, L., Bainton, D.F., 1989, GMP-140, a platelet alpha-granule membrane protein, is also synthetized by vascular endothelial cells and is localized in Weibel-Palade bodies. J. *Clin. Invest.* 84: 92-99.

Picker, L.J., 1992, Mechanisms of lymphocyte homing. *Curr. Opin. Imrnunol.* 4:244-286.

Siegelman, M.H. van de Rijn, M. Weissman, I.L. 1989, Mouse lymph node homing receptor cDNA clone encodes a glycoprotein revealing tandem interaction domains. *Science,* 243: 1165-1172;

Springer, T.A., 1990, Adhesion receptors in the immune system. *Nature,* 346:425-434.

USE OF DIFFERENT CYTOMETRIC TECHNIQUES TO STUDY THE CYTOTOXIC INTERACTION BETWEEN HUMAN NATURAL KILLER CELLS AND K562 TARGET CELLS

Katarina Radošević, Bart G. de Grooth and Jan Greve
University of Twente
Dept. of Applied Physics, Applied Optics Group
P.O. Box 217
7500 AE Enschede
The Netherlands

Introduction

The studies described in this manuscript have been aimed to obtain a better insight into *the cytotoxic interaction* between human Natural Killer (NK) cells and K562 target cells[1]. In order to obtain information about crucial processes leading to target cell death, the processes taking place in *the target cell* during a killer cell attack were studied using *different optical techniques*. Furthermore, *flow cytometric methods for determination of the phenotype and the cytotoxic activity of killer cells* were developed.

Cytotoxic cells, cytotoxic lymphocytes (CTL) and Natural Killer (NK) cells, are powerful mediators of cellular immunity, playing an important role in the control of the organism integrity. They can attack and destroy numerous types of cells in order to protect the organism against transformed cells or infecting agents.
The cytotoxic activity, performed by both NK cells and CTL, can generally be divided in a few stages (illustrated in Figure 1).
Binding of a killer cell to a target cell (conjugate formation) is a prerequisite for the recognition of a target cell, i.e. activation of the cytotoxic cell for killing, to take place. Upon binding, *programming* of the target cell for lysis takes place. This stage is temperature sensitive

---

[1] An extended bibliography on the subjects described in this paper can be found in the corresponding references.

NATO ASI Series, Vol. H 95
Flow and Image Cytometry
Edited by Alain Jacquemin-Sablon

(optimum at 37^0C) and requires Ca^{2+}. In the last, killer cell independent, stage of the cytotoxic process cytolysis of the target cells takes place.

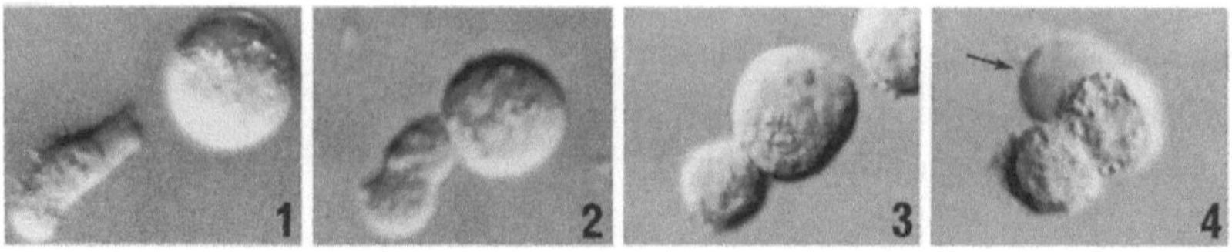

Figure 1. *Cytotoxic process as imaged using Nomarski optics.*
1. *NK cell (bottom left) moves towards the K562 cell (top right).*
2. *Binding (i.e. conjugate formation).*
3. *Programming of the target cell for lysis.*
4. *Cytolysis of the target cell, accompanied by the bleb formation (arrow).*

Attempts to resolve the killing mechanism have been the subject of numerous studies. Nowadays it is generally accepted that each type of cytotoxic cell may use more than one killing mechanism (or better may employ different killing molecules) to destroy a target cell.
Attacks on two major organelles of target cells have been identified. Damages of the cell *membrane* (primarily by perforin), resembling that caused by complement or some bacterial toxins, may lead to colloid osmosis which is morphologically identified as necrosis. In *DNA*, as another target, breaks may be induced that are of the single- or double-stranded type, depending on the type of target cell. It is not clear whether killer cell enzymes are *directly* involved in causing this DNA damage or whether they only *activate* a self-destruction path in a target cell, as in apoptosis.

Flow Cytometric Studies

Conjugate Formation and Immunophenotyping

The capability of a cytotoxic cell to form a conjugate with a target cell is a prerequisite for the killing process to take place. It is important, therefore, to have a technique that enables detection and quantification of the conjugates. Since distinct populations of peripheral blood cells can form conjugates with the NK-sensitive targets, the technique should also enable the determination of the phenotype of the conjugate forming cells.
A flow cytometric method for the simultaneous quantification of conjugates and immunophenotyping of conjugate forming cells has been developed in our lab (Radošević et al., 1993b). The method uses three fluorescent probes. One probe (F-18) is used to label PBL

prior to incubation with K562 cells. After incubation the cells are treated with monoclonal antibodies labeled with phycoerythrin and Red613, respectively. The combination of F-18 fluorescence and light scattering enables identification and quantification of conjugates while the fluorescence of the monoclonal antibodies provides information about the phenotype of the conjugate forming cells.

The identification of cell populations is demonstrated in Figure 2.

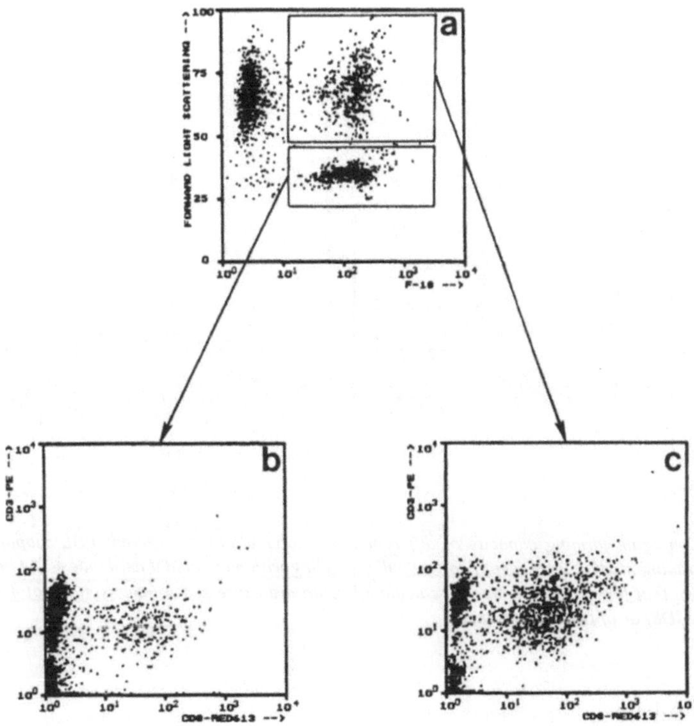

Figure 2. *Scatter plots showing identification of cell populations. F-18 labeled PBL were incubated with K562 cells to form conjugates and the sample was stained with monoclonal antibodies (anti-CD3-PE + anti-CD8-biotin/streptavidin-Red613). Single cells and conjugates were identified on the basis of F-18 fluorescence and light scattering signal (a). Phenotypic identification could be done for both non-conjugated PBL (b) and conjugated PBL (c). In each plot 2 500 cells are shown.*

Validity of the method was tested using monoclonal antibodies that identify NK cells (anti-CD16, anti-CD56) and T lymphocytes (anti-CD3), both cytotoxic (anti-CD8) and helper cells (anti-CD4). In agreement with the expectations, the results showed that CD8+ PBL, are predominant among the cells that form conjugates with K562 cells. These subtypes include both NK cells (CD16+, CD56+) and CTL (CD3+). Analysis of conjugate forming capacity (i.e. the percentage of a subpopulation that forms a conjugate) of PBL subtypes revealed the CD56+CD8+ (mostly NK cells) population as the type with the highest conjugate forming capacity (Figure 3). Our results demonstrated that the method can be used for the simultaneous quantification and immunophenotyping of the conjugates.

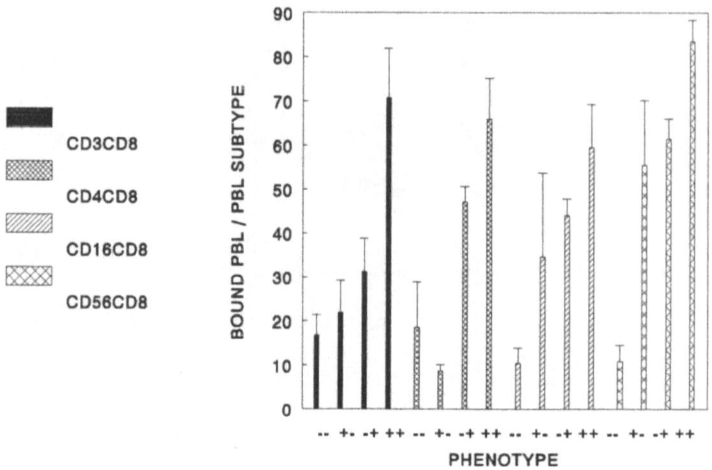

Figure 3. *Conjugate forming capacity of different PBL subpopulations. Different PBL subpopulations in a conjugate forming mixture were identified by double labeling with monoclonal antibodies. The percentage of a subpopulation that forms a conjugate was calculated. Data are expressed as means ± s.d. of 4 subjects (anti-CD3 + anti-CD8) or of two subjects (rest).*

Cytotoxicity Assay

Spontaneous cytotoxic activity in vitro against a variety of cells is the most prominent functional characteristic of NK cells. It is, therefore, desirable to have a method to determine their cytotoxic activity. The most commonly used assay for this purpose is the ^{51}Cr-release assay.

We have developed *a flow cytometric cytotoxicity assay* (Radošević et al., 1990). The principle of the method is illustrated in Figure 4. Target cells are labeled with membrane probe F-18 and mixed with killer cells in the desired ratio. After incubation at 37^{0}C for the desired period of time (usually 2 hours), propidium iodide (PI) is added and the cell suspension is analyzed using the flow cytometer. The relative number of PI positive F-18-labeled cells (i.e. dead target cells) is a measure of the cytotoxic activity.

F-18 was tested and it fulfilled all requirements for use in the cytotoxicity assay. The probe is not toxic and does not decrease the cytotoxic activity of NK cells. It is also stable (exchange between labeled and non-labeled cells is negligible in a period of at least 4 hours at 37^{0}C) and it remains in the membrane of the target cells even after they have been killed. A clear distinction between unlabeled killer cells and labeled target cells is obtained.

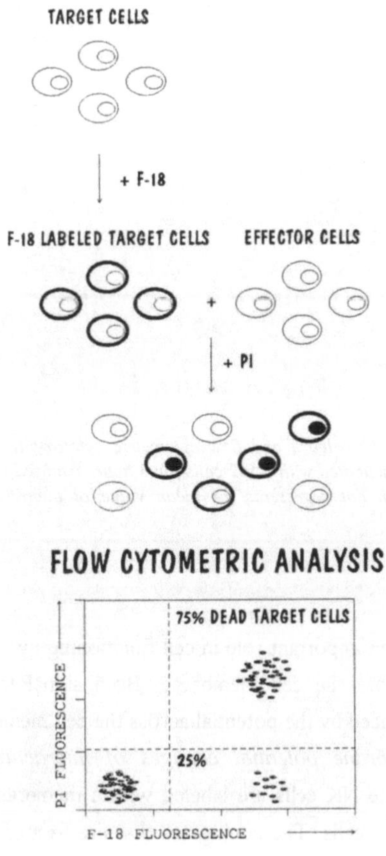

Figure 4. *The principle of the flow cytometric cytotoxicity assay.*

As can be seen from Figure 5, there is a good correlation between the flow cytometric and the ^{51}Cr-release assay.

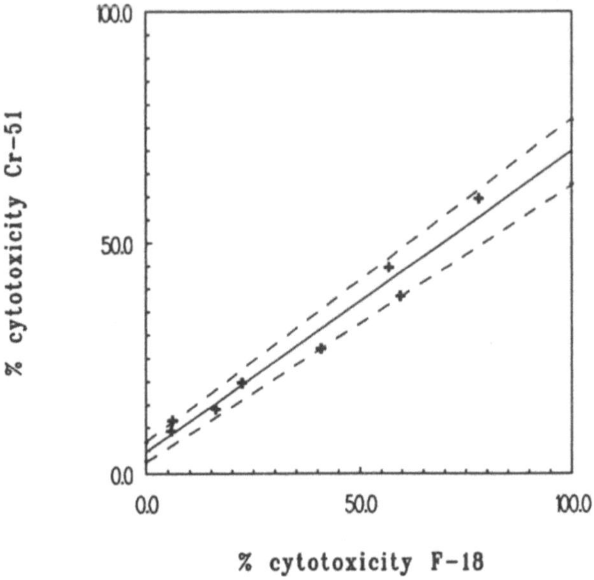

Figure 5. *Correlation between ^{51}Cr-release and flow cytometric cytotoxicity assays. Effector cells (PBL) isolated from three donors were incubated with K562 cells for 3 hours (ratios 25:1, 50:1 for all three donors, ratio 75:1 for two donors). Each dot represents the mean value of the results of three measurements. Correlation coefficient = 0.98.*

<u>Membrane Potential</u>

The membrane potential has an important role in cell functioning by controlling, either directly or indirectly, ion fluxes across the cell membrane. Both signal transduction and osmotic balance of the cells are influenced by the potential across the cell membrane.

We have studied *the membrane potential changes of interacting NK and K562 cells* (Radošević at al., 1993a). The NK cells are labeled with a membrane probe (TR-18 or DiI) prior to incubation with K562 cells. The conjugates that are formed can be identified on the basis of the membrane probe fluorescence and light scattering signals. The membrane potential of the cells is monitored using $DiBAC_4(3)$, an anionic probe. Upon depolarization, when the inside of the cell becomes more positively charged, more probe enters the cell resulting in an increase of cellular fluorescence. Upon hyperpolarization, the probe is extruded from the cell which results in a decrease of fluorescence.

As can be seen from Figure 6, we detected a depolarization (i.e. an increase in the fluorescence intensity of $DiBAC_4(3)$ labeled cells) of a fraction of the conjugates.

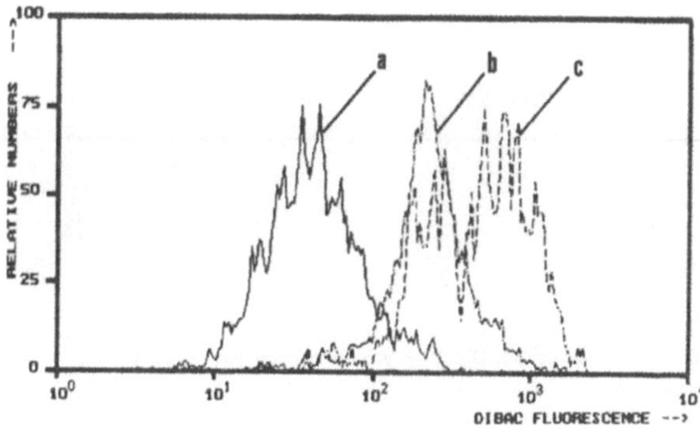

Figure 6. *DiBAC$_4$(3) fluorescence histograms of a mixture of K562 and NK cells obtained immediately after conjugate formation. a, single NK cells; b, single K562 cells; c, conjugates.*

The DiBAC$_4$(3) fluorescence signal of the conjugate is the sum of the DiBAC$_4$(3) fluorescence signals of the cells that form the conjugate. Therefore, using standard flow cytometry it is not possible to resolve in which of the conjugate forming cells the change in the membrane potential occurs. Using the slit-scan analysis technique that had been developed in our lab (Bakker Schut et al., 1993) we could determine the membrane potential of each cell in a conjugate separately. The principle of the technique is illustrated in Figure 7.

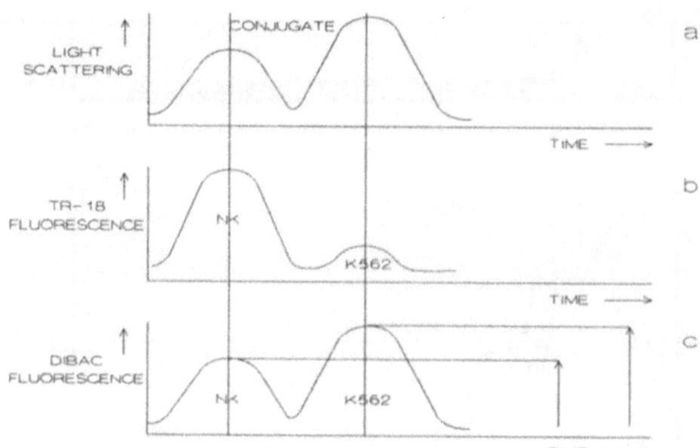

Figure 7. *Principle of the slit-scan technique used to monitor DiBAC$_4$(3) fluorescence of each cell in a conjugate separately. The conjugates were detected on the basis of the pulse form of the scattering signal (a). Cells in the conjugate were identified on the basis of TR-18 fluorescence (b) and the intensity of their DiBAC4(3) fluorescence was determined (c).*

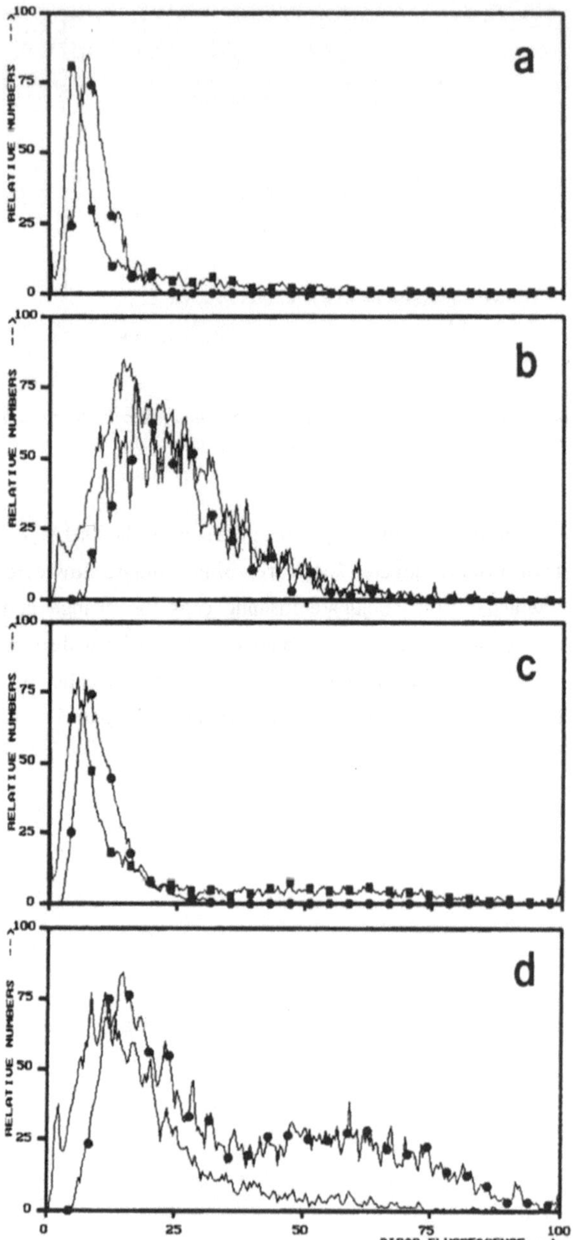

Figure 8. *$DiBAC_4(3)$ fluorescence histograms of conjugated and single NK and K562 cells obtained with the slit-scan flow cytometer. TR-18 labeled NK cells were incubated with K562 cells for 0 min (a,b) and 10 min (c,d) at 37^0C, stained with $DiBAC_4(3)$ and analyzed by the slit-scan flow cytometer. a c, histograms of NK cells (■——■, single; •——•, conjugated); b,d, histograms of K562 cells (——, single; •——•, conjugated).*

The depolarization of conjugates observed with the standard flow cytometer is primarily due to depolarization of *K562 target cells*. This can be seen from Figure 8, where $DiBAC_4(3)$ fluorescence histograms of conjugated and single NK and K562 cells obtained with the slit-scan flow cytometer are shown.

The membrane of most of the depolarized cells still remained impermeable to propidium iodide indicating that depolarization represents an early sign of cell damage.

This study demonstrates that it is possible to investigate membrane potential of interacting cells using fluorescent probes and flow cytometry.

Microscopic Studies

As demonstrated, flow cytometry can provide variety of useful information about the interacting cells. In some cases, however, it is not possible to obtain desired information. If one wants to get detailed spatial and temporal resolution at the single cell level or morphological information, flow cytometry may not be satisfactory. In such cases microscopic techniques are very useful.

Intracellular Calcium Concentration and Intracellular pH

The cellular physiology is greatly affected by the processes that lead to cell damage. In order to get a better insight into the processes associated with target cell death, we have made an attempt to characterize *the changes that occur in target cell intracellular calcium concentration* ($[Ca^{2+}]_{in}$) *and intracellular pH* ($[pH]_{in}$) during an NK cell attack (Radošević et al.,1995a). We have done so by performing quantitative fluorescence microscopy of the cells labeled with ratio probes specific for $[Ca^{2+}]_{in}$ (Fura-2-AM) and $[pH]_{in}$ (BCECF-AM). A change in the fluorescence ratio of these probes indicates a change in the physiological parameter. We used a leakage of the probes out of the cell as an indicator of membrane damage. Complete loss of the probes was interpreted as a sign of cell death.

Our results showed that two types of target cell response to the NK cell attack can be distinguished. The cells either die immediately (further referred to as *dying* cells), due to the complete breakdown of the membrane impermeability, or the initial membrane damage is repaired and cells "escape" immediate death (*transient* cells). During both responses an increase of $[Ca^{2+}]_{in}$ takes place in the target cells. This increase is illustrated in Figure 9, middle column. The increase of $[Ca^{2+}]_{in}$ is usually accompanied/followed by a leakage of Fura-2 out of the cell indicating an increase in the permeability of the cell membrane, i.e. cell damage (Figure 9, right column).

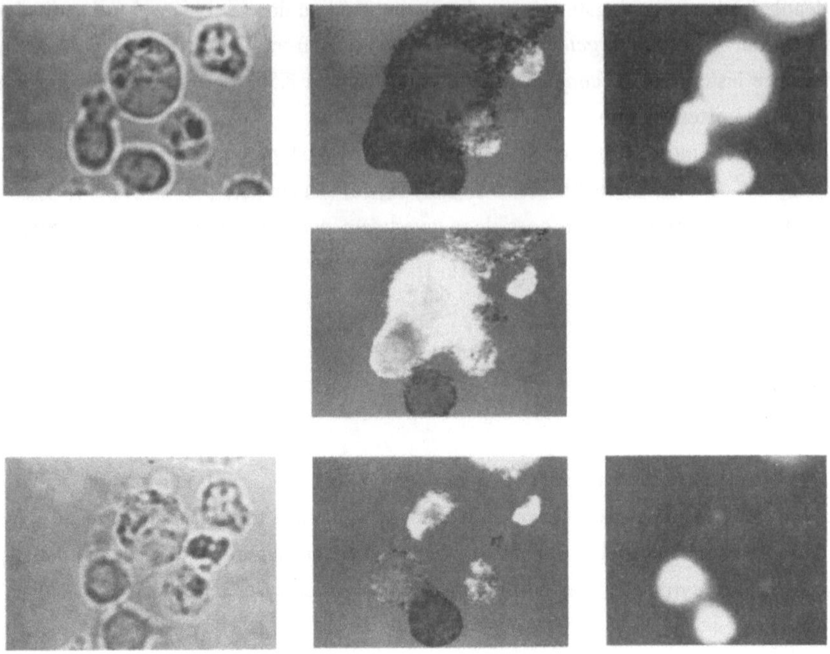

Figure 9. *Increase of $[Ca^{2+}]_{in}$ in the K562 cell during the NK cell attack.*
Left column: bright field image (middle, K562 cell; bottom left, NK cell).
Middle column: ratio of Fura-2 (i.e. $[Ca^{2+}]_{in}$).
Right column: fluorescence upon 340 nm excitation of Fura-2.
The brightness intensity corresponds to fluorescence intensity and ratio level. Top row, before the killing process (i.e. incubation in calcium-free medium); middle row, 2 min incubation in calcium rich medium; bottom row, 6 min incubation in calcium-rich medium.

Figure 10 shows the average changes in $[Ca^{2+}]_{in}$ as a function of time for dying (closed symbols) and transient (open symbols) target cells. As can be seen from the figure, in the cells that die immediately $[Ca^{2+}]_{in}$ reaches higher levels than in the cells that restore the initial damage.

In addition to an increase of $[Ca^{2+}]_{in}$, both dying and transient cells undergo changes in $[pH]_{in}$. The direction of the change (acidification or alkalization) as well as the level of the change depend on extracellular pH. The changes in $[pH]_{in}$, thus, primarily reflect an equilibration with extracellular pH due to the increased membrane permeability.

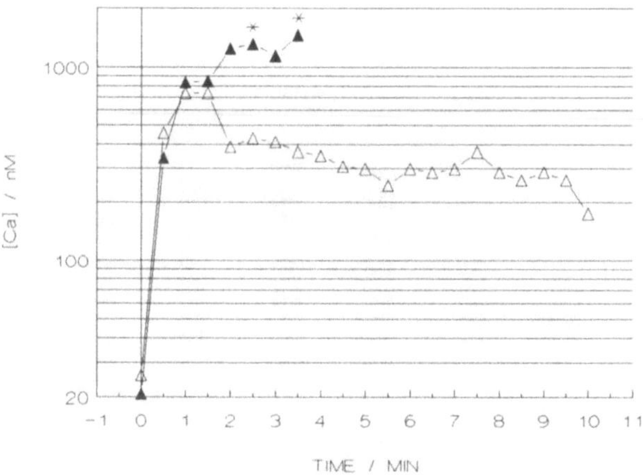

Figure 10. *Changes in $[Ca^{2+}]_{in}$ of dying (closed symbols) and transient (open symbols) K562 cells as a function of time. The average $[Ca^{2+}]_{in}$ of 15 transient cells is shown. For dying cells the number of cells available for averaging decreased gradually from 16 (first 2 points) to 4 (last 2 points) due to cell death (i.e. complete leakage of the dye out of the cells). Time 0 min indicates the last measurement before the increase in $[Ca^{2+}]_{in}$ was detected. (* - significantly different from the transient cells at the same time point; $p<0.05$).*

Accumulation of a fluorescent probe in certain cellular compartments is usually not desirable. However, this phenomenon can be used as a morphological marker. Fura-2-AM may accumulate in the granules of killer cells in unhydrolyzed form. As will be demonstrated during the lecture, this nonuniform Fura-2 fluorescence distribution can indirectly be used to demonstrate an involvement of the killer cell granules in the cytotoxic process.

Actin Cytoskeleton

The cytoskeleton may play an important role during the cytotoxic process. Intact microfilaments seem to be important for conjugate formation and postlytic recycling of NK cells. What is happening to microfilaments of cells that are attacked by NK cells has not been investigated very extensively. Therefore we have studied ***the organization of F-actin in K562 target cells during the cytotoxic process***, using quantitative fluorescence microscopy (QFM) and confocal laser scanning microscopy (CLSM) (Radošević et al., 1994b). For the quantification experiments we have used rhodamine phalloidin (exhibits greatest fluorescence enhancement of all phalloidin derivatives upon binding to F-actin), while CLSM experiments were performed using rhodamine phalloidin or BODIPY 581/591 phalloidin (the fluorescent derivatives of phalloidin that exhibit a low photobleaching rate). Due to more intense labeling with phalloidin, NK cells could clearly be distinguished from K562 cells.

Using QFM we detected an increase in the F-actin content in both conjugated NK and conjugated K562 cells during the cytotoxic process (Figure 11).

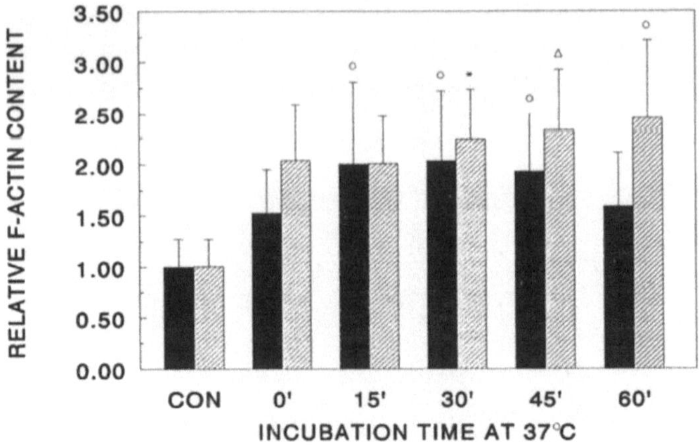

Figure 11. *The relative content of F-actin in conjugated cells. The mean intensities of rhodamine phalloidin fluorescence of conjugated NK and conjugated K562 cells were normalized with respect to the mean intensities of rhodamine phalloidin fluorescence of control NK and control K562 cells, respectively, for each incubation time. CON represents the mean intensities of rhodamine phalloidin fluorescence of all measured control NK and control K562 cells, respectively. For each incubation time, 75 conjugated and 75 control cells were measured. The error bars represent standard deviations of the measurements. There was an significant increase in F-actin content in the conjugated cells for all incubation times compared to the control cells ($p<0.001$, except for the conjugated K562 cells upon 15 min incubation: $p<0.05$). (°, significantly different form conjugated cells on 0 min incubation, $p<0.001$; *, significantly different from conjugated cells on 0 min incubation, $p<0.01$; Δ, significantly different from conjugated cells on 0 min incubation, $p<0.05$); black bars, conjugated K562 cells; shaded bars, conjugated NK cells.*

The increase in F-actin content in the conjugated K562 cells during the cytotoxic process occurs simultaneously with profound changes in the organization of F-actin, as observed using CLSM (Figure 12). Instead of the poorly filamentous F-actin structures, consisting of the cytoplasmic aggregates of F-actin and F-actin rich microvilli (Fig. 12A), cytoplasmic filamentous structures develop in conjugated K562 cells during the cytotoxic interaction. The filaments are sometimes oriented from the contact place with the NK cell towards the opposite pole of the K562 cell. In the phase of the cytotoxic process where blebs appear on K562 cells, well-defined and loosely arranged F-actin filaments can be observed in the blebs, while in the cytoplasm more tightly packed F-actin filaments are localized (Fig. 12B). Using additional labeling with Hoechst 33258 (DNA probe) and UV laser excitation, actin filaments were occasionally seen to spread across the target cell nucleus forming indentations. We do not know whether filaments deformed nucleus or they were polymerized over already existing nuclear grooves.

The role and the importance of these changes for the cytotoxic process remain to be resolved.

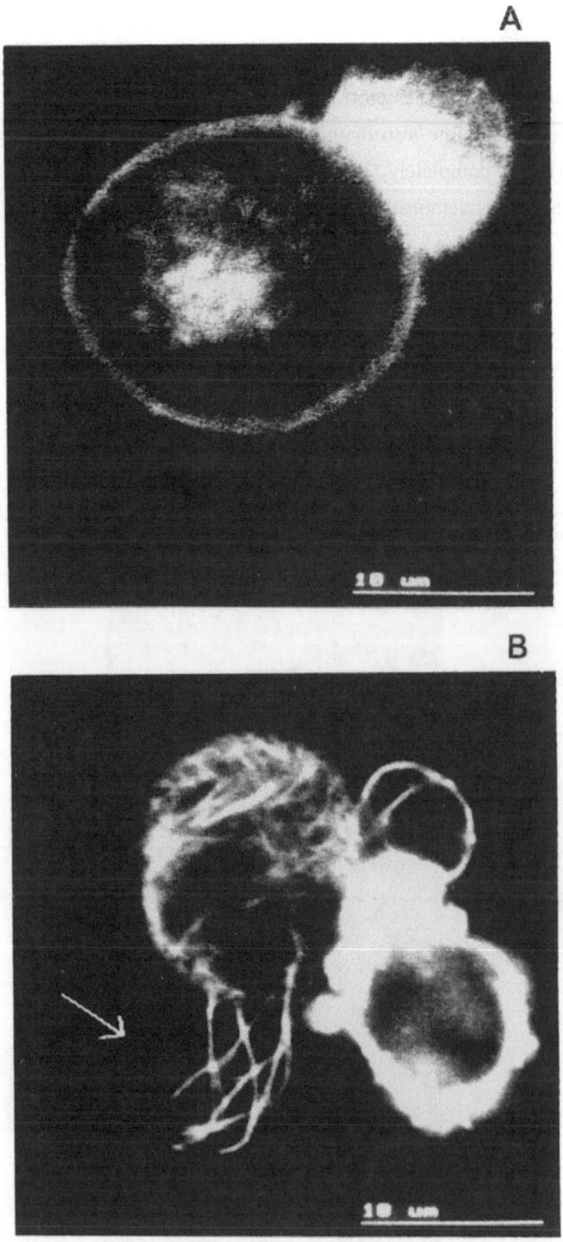

Figure 12. *CLSM images of rhodamine phalloidin labeled F-actin in conjugates.*

A: conjugated K562 cell with the poorly filamentous F-actin structures (top right, NK; bottom left, K562).

B: F-actin filaments in blebs (arrow) of conjugated K562 cell (bottom right, NK; top left, K562).

In-Conjugation

Using the optical sectioning capability of CLSM we frequently observed an interesting phenomenon occurring during the cytotoxic interaction. Namely, ***killer cells*** do not just firmly attach to target cells but some also ***deeply penetrate into the cytoplasm of the target cells***, occasionally entering it completely (Radošević et al., 1995b). This phenomenon, which we named "in-conjugation", is demonstrated in Figure 13.

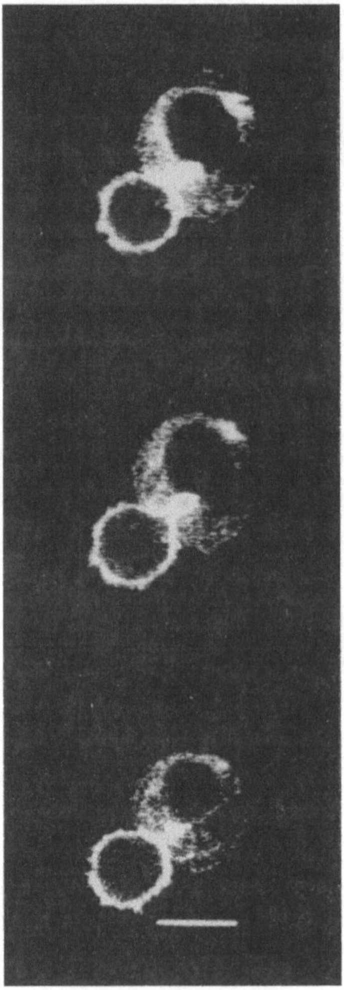

Figure 13. *Three optical sections (1 μm apart) through a rhodamine phalloidin labeled in-conjugate, containing NK cell with a finger like projection. The bar represents 10 μm.*

Using additional labeling with Hoechst 33258 and UV laser excitation, we observed occasionally a deformation of the nucleus of a target cell by the F-actin rich protrusion of a killer cell.

In order to confirm that the in-conjugation phenomenon was not an artefact due to the fixation or labeling procedures we have performed CLSM measurement using living interacting NK and K562 cells. For this purpose we have used *double-double vital labeling method* which enables a morphological CLSM study of living interacting cells (Radošević et al., 1994a). The basic idea of the method is to label each cell type with two dyes with overlapping emission spectra but with different cellular localization. Using dye pairs that differ in their spectral characteristics enables distinction between two cell types. We have labeled NK cells with DiI (membrane) and SNARF-1-AM acetate (cytoplasm) (both are red fluorescing dyes) while K562 cells were labeled with DiO (membrane) and Calcium Green-1-AM (cytoplasm) (these are both green fluorescing dyes). Dye pairs were simultaneously excited using 488 and 568 nm lines of an Ar-Kr laser. Using this labeling method we obtained two fluorescent signals that were distinguishable from each other and strong enough to perform confocal microscopy while the cells stay alive. Using this method we demonstrated that in-conjugation indeed occurs during the cytotoxic interaction.

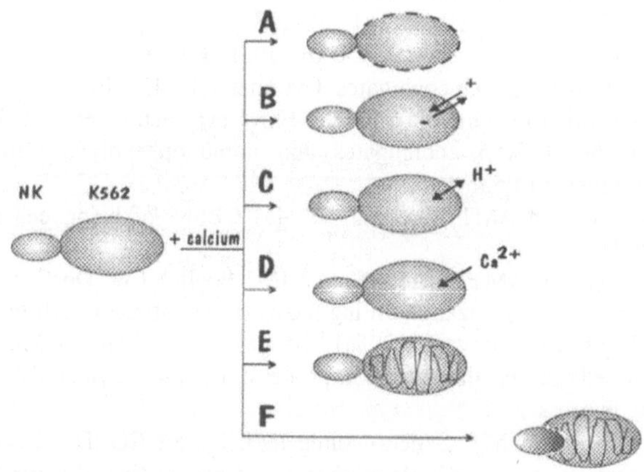

Figure 14. *Overview of the changes detected in a target cell during the cytotoxic process.*

A. increased membrane permeability
B. membrane depolarization
C. changes in $[pH]_{in}$
D. increase in $[Ca^{2+}]_{in}$
E. increase in F-actin
F. in-conjugation

In-conjugation appeared to be dependent on the presence of calcium, actin polymerization and an endonuclease activity. The mechanism and the role of this interesting phenomenon for the cytotoxic process are not resolved.

The changes detected in the target cell are schematically shown in Figure 14.

Conclusion

These studies of the cytotoxic process illustrate that by means of different cytometric techniques the variety of information can be obtained about a process involving interacting cells

Radošević K, Garritsen HSP, Van Graft M, De Grooth BG, Greve J (1990) A simple and sensitive flow cytometric assay for the determination of the cytotoxic activity of human natural killer cells. J Immunol Methods 135:81-89.

Radošević K, Bakker Schut TC, Van Graft M, De Grooth BG, Greve J (1993a) A flow cytometric study of the membrane potential of natural killer and K562 cells during the cytotoxic process. J Immunol Methods 161:119-128.

Bakker Schut TC, Florians A, Radošević K, De Grooth BG, Greve J (1993) Signal processing in slit-scan flow cytometry of cell conjugates. Cytometry 14:459-464.

Radošević K, De Grooth BG, Greve J (1993b) Flow cytometric method for simultaneous detection of lymphocyte-K562 conjugates and immunophenotyping of the conjugate forming cells. Cytometry 14:535-540.

Radošević K, Van Leeuwen AMT, Greve J (1994a) CLSM of living cell conjugates. Eur Microsc Anal 28:41.

Radošević K, Van Leeuwen AMT, Segers-Nolten IMJ, Figdor CG, De Grooth BG, Greve J (1994b) Changes in actin organization during the cytotoxic process. Cytometry 15:320-326.

Radošević K, De Grooth BG, Greve J (1995a) Changes in intracellular calcium concentration and pH of target cells during the cytotoxic process: a quantitative study at the single cell level. Cytometry, in press.

Radošević K, Van Leeuwen AMT, Segers-Nolten IMJ, Figdor CG, De Grooth BG, Greve J (1995b) The occurrence and a possible mechanism of penetration of natural killer cells into K562 target cells during the cytotoxic interaction. Cytometry, in press.

In vitro Reconstitution of Early Endosome Membrane Dynamics

Jean-Pierre Gorvel and Zohair Mishal [1]
Centre d'Immunologie INSERM-CNRS de Marseille-Luminy
Case 906
13288 Marseille Cedex 9
France

KEYWORDS: endocytosis/ endosome/ flow cytometry/ in vitro assay/ membrane fusion/ membrane fission

ABSTRACT:
Flow cytometry was used for the characterization of the molecular mechanisms involved in early endosome membrane fusion and. The roles of rab4 and rab5 GTPases were investigated. The membrane fusion between individual early endosomes was strongly stimulated by rab5 whereas rab4 had no effect. In contrast, membrane fission was strongly stimulated by rab4Q67L, a form analogous to the GAP insensitive p21-rasQ61L oncogenic mutant. Membrane fission was GTPγS sensitive and NEM insensitive.

INTRODUCTION:

In vitro reconstitution assays have provided important insights into the analysis of biosynthetic and endocytic mechanisms in membrane transport (Balch, 1989; Goda and Pfeffer, 1989; Gruenberg and Clague, 1992; Rothman and Orci, 1992; Schekman, 1992; Ferro-Novick and Jahn, 1994; Latterich and Schekman, 1994). It has been suggested that intracellular membrane equilibrium is controlled by homotypic fusion and fission events (Gruenberg and Howell, 1989; Rothman and Warren, 1994). In the biosynthetic pathway, homotypic fusion assays that reconstitute the nuclear envelope formation (Wilson and Newport, 1988; Boman et al., 1992) and the endoplasmic reticulum (ER) membrane fusion (Watkins et al., 1993; Latterich and Schekman, 1994) have been reported. In the endocytic pathway, early endosomes exhibit with each other a high propensity to undergo homotypic fusion (Davey et al., 1985; Woodman and Warren, 1988; Gruenberg and Howell, 1989). This

[1] *CNRS UPS 47 Laboratoire de Cytométrie, 7 rue Guy Moquet 94800 Villejuif, France.*

NATO ASI Series, Vol. H 95
Flow and Image Cytometry
Edited by Alain Jacquemin-Sablon

process is highly specific (Gruenberg et al., 1989; Gorvel et al., 1991; Aniento et al., 1993), involves NSF (Diaz et al., 1989), and is controlled by GTPases (Gorvel et al., 1991; Colombo et al., 1992; Lenhard et al., 1992). Two small GTPases have been described which are associated with early endosomes : rab5 and rab4 (Chavrier et al., 1990; van der Sluijs, 1991, 1992). Overexpression of wild-type rab5 (rab5wt) or rab5Q79L, a mutant form locked in its active GTP-bound conformation causes the formation of large early endosomal structures (Bucci et al., 1992; Stenmark et al., 1994) and increases early endosome fusion efficiency *in vitro* (Gorvel et al., 1991; Stenmark et al., 1994). In contrast, rab5 mutants with single point mutations in the GTP/GDP binding motives resulting in impaired GTP-binding (rab5N133I and S34N) inhibit early endosome fusion *in vitro* (Gorvel et al., 1991; Stenmark et al., 1994), endocytosis *in vivo* (Bucci et al., 1992; Stenmark et al., 1994) and lead to the fragmentation of early endosomes (Bucci et al., 1992; Parton et al., 1992a). Although the role of rab4 has not yet been investigated *in vitro* and the precise step controlled by rab4 is still unidentified, it has been suggested that rab4 plays an important role in the sorting or transit of transferrin receptor and fluid phase markers through early endosomes (van der Sluijs et al., 1992).

The high fusion activity of early endosomes observed *in vitro* suggests that early endosomes are highly dynamic *in vivo* (Gruenberg et al., 1989; Parton et al., 1992a). Morphologically, electron-microscopic observations revealed that early endosomes are organized as a tubulo-vesicular network in several cell types (Tooze and Hollinshead; 1991) and a dynamic endosomal network could be visualised by time-lapse video microscopy (Hopkins et al., 1990). It has been suggested that this dynamic network depends on the capacity of individual elements of early endosomes to undergo fusion and fission events *in vivo* (Gruenberg and Howell, 1989).

In this work, we used a flow cytometry assay for a direct study of early endosome membrane dynamics. We show that early endosomes undergo homotypic fusion reactions efficiently stimulated by rab5. After fusion, early endosomes undergo fission events which are GTPγS sensitive and NEM insensitive and highly stimulated by rab4 and rab4Q67L. From these data, we argue that early endosomes are highly dynamic organelles governed by fusion and fission events, the control of which is ensured by at least two GTPases : rab5 and rab4.

RESULTS:

The three color cytometry assay.

BSA-fluorescein and BSA-rhodamine were separately internalized for 5 min at 37°C in BHK cells. Early endosome fractions (EE) were prepared according to Gorvel et al., (1991) and used in the cell free assay as described in Material and methods. After various incubation times at 37°C, reactions were stopped at 4°C and EE were labeled at 4°C with anti-rab5 or anti-rab4 antibodies and detected with an streptavidin-allophycocyanin (APC) system. The amount of endosomes containing the three markers (BSA-fluorescein, BSA-rhodamine and allophycocyanin-stained rab5 or rab4) was expressed as a percentage of the total APC positive endosomal population. After 15 min at 37°C, approximately 45% of rab5 or rab4 positive endosomes contained both BSA-fluorescein and BSA-rhodamine compared to about 10% observed at 4°C. Figure 1 shows that the appearance of the three color-positive endosome subpopulation fulfills the criteria of early endosome fusion. It is NEM, GTPγS, ATP and temperature-dependent as previously described (Gruenberg et al., 1989; Gorvel et al., 1991).

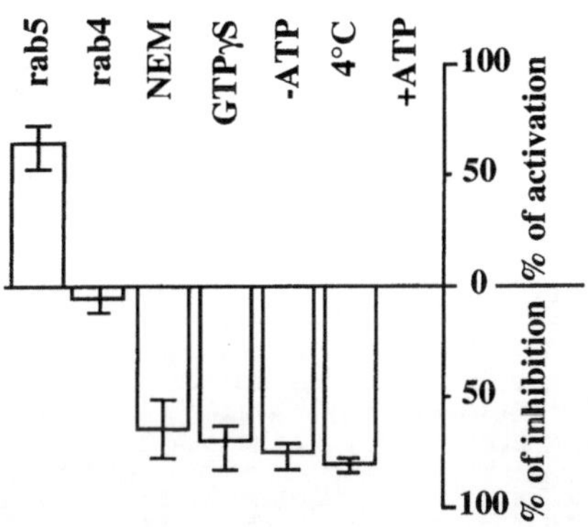

Figure 1: Triple labeled EE population analyzed by flow cytometry. The reference assay, termed (0) in ordinates corresponds to the assay performed in the presence of an ATP regenerating system (+ATP). Stimulatory or inhibitory effects were analyzed by adding in the assay 1.5 μg of GSTrab5 or GSTrab4 (rab5, rab4); 1mM N-ethylmaleimide (NEM); apyrase (-ATP), 10 μM GTPγS. The quantification of triple labeled EE was determined after 15 min at 37°C or 10 min at 4°C (when indicated). Elite flow cytometer settings were as in Materials and methods. 33,000 elements were analyzed in each sample.

Since the number of endosomes labeled by anti-rab4 was similar to those labeled with anti-rab5-antibodies, endosomes labeled by anti-rab5 specific antibodies allowed us to study the effects of GSTrab4 and conversely, rab4 antibodies were used when the effects of GSTrab5 were studied. Rab4 or rab5 bacterially-expressed proteins as fusion proteins with the glutathion-S-tranferase (GST) were added in the assay (Figure 1). As previously described, rab5 stimulates the early endosome fusion activity (Gorvel et al., 1991; Aniento et al., 1993; Stenmark et al., 1994). In contrast, rab4 had no effect suggesting that it is not an essential element for the control of early endosome fusion *in vitro*.

GTPase defective rab4Q67L mutant prevents 3-7 μm EE formation and induces the generation of 1-3 μm triple labeled EE.

Statistical analyses of the number of triple-labeled 1-3 µm and 3-7 µm EE subpopulations were performed by flow cytometry. The flow cytometer was calibrated for the best separation of 0.95 µm, 2 µm and 5 µm fluorescent beads both in the forward (FALS) and right angle light scatter (90LS) modes. Gates were taken to analyse endosomal subpopulations ranging between 1 and 3 µm (referred as 1-3 µm EE) and between 3 and 7 µm (referred as 3-7 µm EE). Figure 2 shows results obtained when we analyzed the 3-7 µm EE subpopulation.

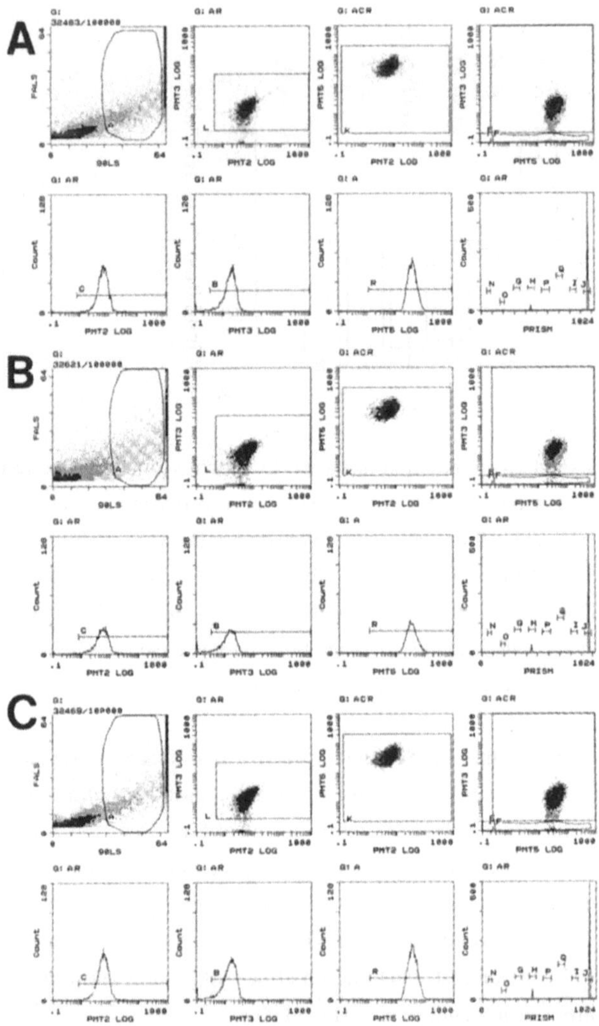

Figure 2: Effects of GSTrab4CT3 and GSTrab4Q67L on the 3-7 µm EE subpopulation. The cell free assay was performed for 10 min at 4°C followed by 15 min at 37°C in the presence

of an ATP regenerating system (A) supplemented with 1.5 μg of GSTrab4Q67L (B) or GSTrab4CT3 (C). A gate was chosen in the light scattered mode to select 3-7 μm EE (A). Elite flow cytometer settings were as in Materials and methods. FITC (PMT2): fluorescein channel; TRITC (PMT3): rhodamine channel; APC (PMT5): Allophycocyanin channel. 90LS : right angle light scatter; FALS : forward angle light scatter. Gates (C, B and R) correspond to the specific fluorescence intensities detected in the FITC, TRITC and APC channels, respectively.

From the different histograms, the PRISM function of the Elite flow cytometer (see Material and Methods) indicated that 90% of APC-labeled 3-7 μm EE contained the three dyes. At 4°C, a small number of labeled 3-7 μ m EE were detected in the three channels. Histograms were plurimodal with broad fluorescent peaks suggesting the presence of particle aggregates as already observed by fluorescence anisotropy (not shown). In contrast, after 15 min at 37°C, the number of 3-7 μm EE had significantly increased (Figure 2A). The effects of rab4 mutants expressed as GST fusion proteins on the formation of 3-7 μm EE were then analyzed. The GSTrab4Q67L mutant which was generated by a single point mutation in the second GTP binding site determining a very low rate of GTP hydrolysis gave a strong inhibitory effect on the number of generated 3-7 μm EE (Figure 2B). The GTPase defective GSTrab4CT3 mutant which lacks the isoprenylation site and cannot attach to the early endosomal membrane had no effect on the generation of 3-7 μm EE (Figure 2C) as well as the GST alone (data not shown). The fact that rab4 does not affect significantly early endosome fusion (Figure 1) and that rab4Q67L inhibits the formation of triple labeled 3-7 μm EE prompted us to carry out a comparative analysis of rab4, rab4Q67L, rab4CT3 and rab5 effects on both 1-3 μm and 3-7 μm EE subpopulations. Quantifications of the number of triple EE was determined by using the PRISM function. Results are shown in Figure 3. After 15 min at 37°C, rab5 stimulated the formation of a large number of 3-7 μm EE. In contrast, the 3-7 μm EE formation was inhibited by rab4 and strongly inhibited by rab4Q67L. Consequently, the number of triple labeled 1-3 μm EE increased (not shown).
After 60 min at 37°C, most of triple labeled EE were found in the 1-3 μm EE subpopulation in all conditions except when rab5 was added in the assay (Figure 3). Indeed, most of EE remained at the stage of 3-7 μm EE in the presence of the latter indicating that rab5 locked the conformation of early endosomes in a 3-7 μm size as previously demonstrated in vitro and in vivo (Gorvel et al., 1991; Bucci et al., 1992). In contrast, rab4 and rab4Q67L had an opposite effect and increased the 1-3 μm EE subpopulation (not shown).

These results indicate that rab4 does not prevent early endosome fusion since mixing of FITC and TRITC conjugates happens but rapidly stimulates the generation of 1-3 μm triple labeled endosomes. No effect was observed with the rab4CT3 mutant indicating that rab4 function requires its attachment to the endosomal membrane.

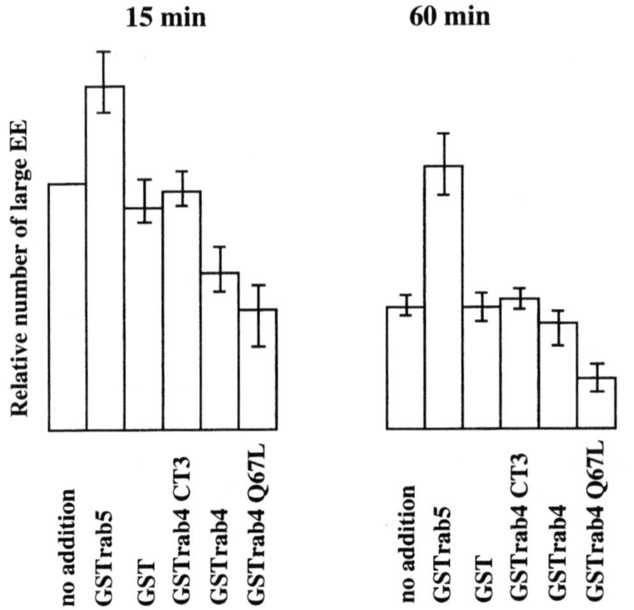

Figure 3: Quantitative analysis of the % of the 3-7 μm EE subpopulation. Statistical analyses were performed by using the PRISM function displayed by the Elite cell sorter. Cell free assays were performed for 15 min or 60 min at 37°C in the presence of an ATP regenerating system (no addition) supplemented with 1.5 μg of GSTrab5, GST, GSTrab4CT3, GSTrab4, GSTrab4Q67L. 33,000 elements were analyzed for each sample. In control experiments (no addition), the 100 reference values represent 12,000 3-7 μm EE detected after 15 min at 37°C, respectively.

Rab4 controls early endosome fission.

We then studied the effects of rab4, rab4CT3, rab4Q67L, GTPγS, NEM and GST when added after 15 min of incubation at 37°C to allow the accumulation of 3-7 μ m fused EE. The kinetics of appearance of both 1-3 μm and 3-7 μm EE subpopulations from 15 up to 60 min were determined (Figure 4).

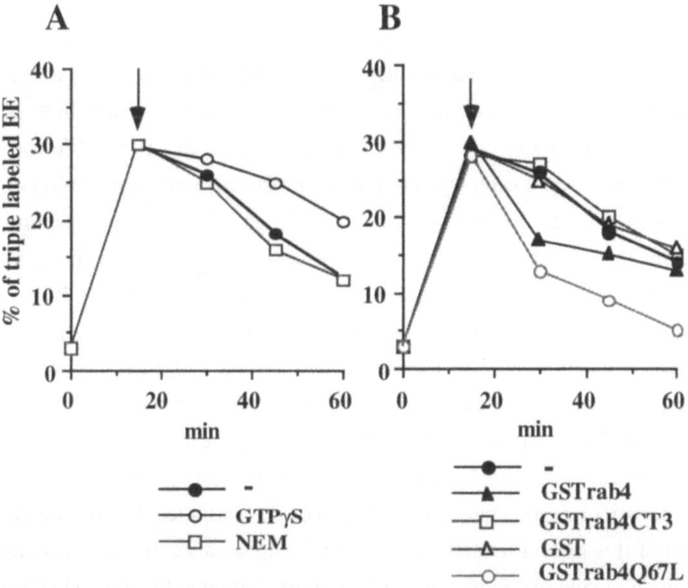

Figure 4: Kinetics of early endosome fusion and fission. % of triple labeled EE were determined from the 3-7 μm EE subpopulation. Samples were analyzed after 10 min at 4°C, 15, 30, 45 and 60 min at 37°C. Cell free assays were done in the presence of an ATP regenerating system (-) supplemented with 1.5 μg of GST, GSTrab4CT3, GSTrab4, GSTrab4Q67L in (B) or with 1mM N-ethylmaleimide (NEM) or with 10 μM GTPγS added after 15 min at 37°C in (A).

We first tested whether or not there was a possible involvement of NSF in early endosome membrane fission by adding N-ethylmaleimide (NEM) after 15 min incubation. Figure 4A shows that in contrast to GTPγS, NEM had no effect. These results suggest that NSF is not involved in fission. In contrast, rab4 and rab4Q67L induced a rapid decrease in the number of 3-7 μm EE whereas rab4CT3 and GST had no effect (Figure 4B). This process was correlated with a stimulation in the appearance of 1-3 μm EE (not shown).

Altogether, these results strongly suggest that the decrease in the number of fused triple labeled 3-7 μm EE immediatly followed by an increase of triple labeled 1-3 μm EE can be interpretated as an EE membrane fission mechanism. Moreover, this step can be blocked by adding rab5 in the assay. Finally, our data strongly favor the role of rab4 in the control of early endosome fission in vitro.

DISCUSSION:

The results presented in this paper show that early endosome membrane dynamics may be dissected in vitro into fusion and fission steps. A three colors cytometry assay allowed us to address the role of two rab proteins in this dynamics : rab4 and rab5 both associated with early endosomes (Chavrier et al., 1990; van der Sluijs et al., 1991). We showed that rab5 controls early endosome fusion whereas rab4 controls early endosome fission *in vitro.*

Extensive studies led to the conclusions that rab5 controls the internalization of solutes and ligands from the plasma membrane towards early endosomes and the homotypic fusion between individual early endosome elements (Gorvel et al., 1991; Bucci et al., 1992; Stenmark et al., 1994). The role of rab5 in early endosome homotypic fusion has been documented by using biochemical assays (Gorvel et al., 1991; Link et al., 1993) and those results were correlated to the electron-microscopic observations showing that overexpression of rab5wt transfected in BHK cells leads to the specific increase of the size of early endosomes (Bucci et al., 1992; Parton et al., 1992a). The effects of rab5Q79L (a GTPase defective form) in early endocytosis has been investigated by Stenmark et al., 1994. Expression of this mutant in BHK cells led to the appearance of structures larger than those observed upon rab5wt overexpression. The authors suggested that rab5:GTP is required prior to membrane fusion. Here, by using flow , we confirm that rab5 induces *in vitro* an increase of the early endosome size and that most of early endosomes remain in the stage of large endosomes in the presence of purified exogenous rab5. In contrast, in the absence of rab5, the large endosome population (3-7 μm relative diameter) progressively disappears to give rise to a small early endosomal subpopulation (1-3 μm relative diameter) containing the two internalized fluorescent markers (BSA-fluorescein and BSA-rhodamine). This indicates that small endosomes are generated from 3-7 μm fused endosomes by a membrane fission process. Moreover, when we used rab4 and rab4Q67L in the assay, we observed an increase rate of formation of small triple labeled early endosomes suggesting that as rab5, rab4:GTP may be required prior to membrane fission.

Altogether,.these results suggest that individual elements of the early endosome are connected by fusion and fission events. Membrane fission may be interpretated as an homotypic event (Figure 5). Individual early endosome elements fuse with each other to generate large structures. This step is favored by rab5:GTP. Upon GTP hydrolyzis, fusion occurs and rab5:GDP is released in the cytosol (Stenmark et al., 1994). Then, fission may take place, stimulated by rab4:GTP, leading to the generation of individual early endosomes. The consequence of this sequential fusion-fission process is a mixing of both membrane and content of individual elements. When rab5Q79L or rab4Q67L mutants with reduced GTPase activity are introduced in this system, they can displace the early endosome membrane equilibrium towards large or small structures, respectively.

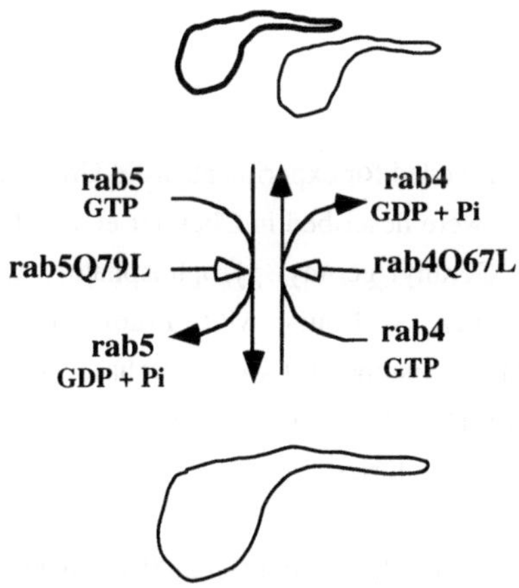

Figure 5: Schematic representation of the role of both rab5 and rab4 in early endosome membrane dynamics. 1-3 µm EE fuse with each other to generate 3-7 µm EE. This step is favored by rab5:GTP. In parallel, 3-7 µm EE undergo membrane fission events controlled by rab4:GTP leading to the generation of 1-3 µm EE. GTPase defective rab5Q79L or rab4Q67L mutants displace the early endosome membrane equilibrium towards 3-7 µm EE or 1-3 µm EE, respectively.

The inhibitory effect of GTPγS on fission suggests that the non-hydrolysable analog exerts a dominant action on rab5 in this system. However, at this stage we cannot rule out the possibility that GTPγS effect may be mediated by other GTPases. For instance, ARF-GTPγS determines an irreversible irreversible recruitment of COP proteins onto Golgi transport vesicles (Orci et al., 1993; Palmer et al., 1993) and ARF proteins are also regulators of endosome fusion (Lenhard et al., 1992). Such homotypic membrane fission events have been reported for peroxisomes biogenesis in the methylotrophic yeast *H. polymorpha* . Under certain growth conditions, peroxisomes undergo a sequence of size-increase followed by fission (Waterham et al., 1993). Small GTP-binding proteins associated to the membrane of this organelle might be involved in this mechanism (Verheyden et al., 1992).

In conclusion, using flow cytometry, we show that the early endosome dynamic network may be characterized by at least two separated steps of membrane fusion and fission controled by rab5 and rab4, respectively. The *in vitro* assay we developed will be useful for the further characterization of molecular machinery involved in the fission mechanism. It would also be interesting to investigate the role of other rab proteins known to be associated with early endosomes (Olkkonen et al., 1993).

MATERIAL AND METHODS:

Cells and Reagents

BHK-21 cells were maintained and seeded for experiments as in Gruenberg et al., (1989). The polyclonal antibodies against rab5 were described in Chavrier et al., (1990) and against rab4 in van der Sluijs et al., (1992). Geranyl geranyl pyrophosphate was purchased from the American Radiolabeled Compagny, St Louis, MO. Fluorescein 5'isothiocyanate or tetramethylrhodamine isothiocyanate (isomer G) were purchased from Molecular Probes, Oregon, USA. Biotin-conjugated F(ab')2 fraction of a goat anti-rabbit IgG and streptavidin-allophycocyanin were from Caltag laboratories.

Cell free assay

BHK-21 cells were scraped from eight 10 cm^2 confluent Petri dishes and early endosome membrane fractions were prepared by flotation gradient after homogenization as previously described by Gorvel et al., (1991). BHK-21 cytosol (12 mg/ml of protein) was prepared as in Gorvel et al., (1991). For each *in vitro* assay, a 100 µl aliquot (40 µg protein) of early endosome fraction was gently mixed with 50 µl of cytosol and adjusted to 12.5 mM HEPES (pH 7.4), 1.5 mM MgOAc, 3mM imidazole, 1mM DTT, 50mM KOAc, and complemented either with 8 µl of an ATP-regenerating system (1:1:1 mixture of 100mM ATP pH 7.0, 800 mM creatine phosphate and 4 mg/ml creatine phosphokinase) or 1.6 µl of apyrase (1200 U/ml) for ATP depletion or 10 µM GTPγS or 1mM NEM or 1.5 µg of GSTrab5, GSTrab4, GSTrab4Q67L, GSTrab4CT3 or GST in the assay in the presence of geranyl geranyl diphosphate (8 µM) as descrided by Lombardi et al., (1993). Samples were incubated for 10 min at 4°C or 15 min at 37°C or 60 min at 37°C. All samples were brought back to ice and the volume was adjusted to 500 µl with cold phosphate saline buffer (PBS) pH 7.4. Detection of rab5 and rab4 was performed by using the affinity-purified rabbit anti-rab5 or anti-rab4 antibodies for 2 h at 4°C at 5 µg/ml.

Three color analysis system

After addition of anti-rab4 or rab5 antibodies in the cell free assay (5µg/ml final in the assay), biotin-conjugated F(ab')2 of goat anti-rabbit IgGs were incubated for 30 min at 4°C at 1 µg/ml final concentration in the assay. This incubation was followed by a allophycocyanin-strepavidin incubation for 30 min at 4°C at 40 µg/ml final concentration in the assay. Then, the samples were diluted to 3 ml with cold PBS and analyzed the Coulter Elite flow cytometer. Statistical analyses of fluorochrome combinations were obtained by using the three color analysis function (PRISM parameter analysis) displayed by the Elite cell sorter. For a three color immunofluorescence, there are eight conceivable phenotypes. The phenotypes are assigned to channels. We were interested in analyzing the percentage of endosomes bearing the three markers (BSA-fluorescein, BSA-rhodamine and rab5 or rab4 detected by the allophycocyanin system (APC)). Channel 960 was referred as the channel detecting the elements containing the three dyes. Statistics analyses were done either by considering the total endosome population (1-7 μm relative diameter) or the population of small endosomes (1-3 μm relative diameter) or the population of bigger endosomes (3-7 μm relative diameter). The flow cytometer was calibrated for the best separation of 0.95 μm, 2 μm and 5 μm fluorescent beads both in the forward and right angle light scatter modes. Upon these conditions, the relative size of rab4 or rab5 positive elements analyzed at 4°C in the absence of cytosol corresponded to a relative size comprised between 1 and 3 μm diameter beads (data not shown). This is in agreement with the size of early endosomes observed by electron-microscopy in BHK cells (Griffiths et al., 1989). A gate was chosen to select the rab5 or rab4 positive endosomes according the APC channel. Without any gate, APC positive EE represented 30 ± 2% of the total element population. By gate addition, they contained 74 ± 15% of rab5 or rab4-positive early endosomes. The arrangement of filters for the detection of fluorescein, rhodamine and allophycocyanin was standard. The cytometer settings were as followed. FSC : 1000 Volt in the linear mode with a gain of 50; SSC : 600 Volts in the linear mode with a gain of 3; fluorescein : 915 Volt in the logarithmic mode with a IGain of 5; rhodamine : 1124 Volt in the logarithmic mode with a IGain of 10; APC : 1100 Volt in the logarithmic mode with a IGain of 10. The laser excitation for fluorescein was 488 nm, 514 nm for rhodamine and 633 nm for APC.

REFERENCES:

Aniento F, Emans N. Griffiths G and Gruenberg J (1993) J. Cell Biol. 123: 1373-1387
Balch WE (1989) J. Biol. Chem. 264: 16965-16968
Boman AL, Delannoy MR and Wilson KL (1992) J. Cell Biol. 116: 281-294
Bucci C, Parton RG, Mather IH, Stunnenberg H, Simons K, Hofflack B and Zerial M (1992) Cell 70: 715-728
Chavrier P, Parton RG, Hauri HP, Simons K and Zerial M (1990) Cell 62: 317-329
Colombo MI, Mayorga LS, Casey PJ and Stahl PD (1992) Science 255: 1695-1697
Davey J, Hurtley SM and Warren G (1985) Cell 43: 643-652
Diaz R, Mayorga L, Weidman PI, Rothman JE and Sthal P (1989) Nature (Lond.) 339: 398-400
Ferro-Novick S and Jahn R (1994) Nature (Lond.) 370: 191-193
Goda Y and Pfeffer SR (1989) FASEB (Fed. Am. Soc. Exp. Biol.) J. 3: 2488-2494
Gorvel JP, Chavrier P, Zerial M and Gruenberg J (1991) Cell 64: 915-925
Griffiths G, Back R and Marsh M (1989) J. Cell Biol. 109: 2703-2720
Gruenberg J and Clague M (1992) Curr. Opin. Cell Biol. 4: 593-599
Gruenberg J and Howell KE (1989) Annu. Rev. Cell Biol. 5: 453-481
Gruenberg J, Griffiths G and Howell KE (1989) J. Cell Biol. 108: 1301-1316
Herman B (1989) Meth. Cell Biol. 30: 219-243
Hopkins CR, Gibson A. Shipman M and Miller K (1990) Nature (Lond.) 346: 335-339
Latterich M and Schekman R (1994) Cell 78: 87-98
Lenhard JM, Kahn RA and Stahl PD (1992) J. Biol. Chem. 267: 13047-13052
Link E, McMahon H, Fischer von Mollard G, Yamasaki S, Niemann H. Südhof TC and Jahn R (1993) J. Biol. Chem. 268: 18423-18426
Lombardi D, Soldati T, Riederer MA, Goda Y, Zerial M and Pfeffer SR (1993) EMBO J. 12: 677-682
Olkkonen VM. Dupree P, Killisch I, Lutcke A, Zerial M and Simons K (1993) J. Cell Sci. 106: 1249-1261
Orci L, Palmer DJ. Amherdt M and Rothman JE (1993) Nature (Lond.) 364: 732-734

Palmer DJ, Helms JB, Beckers CJ, Orci L and Rothman JE (1993) J. Biol. Chem. 268: 12083-12089
Parton RG, Schrotz P, Bucci C and Gruenberg J (1992a) J. Cell Sci. 103: 335-348
Rothman JE and Orci L (1992) Nature (Lond.) 355: 409-415
Rothman JE and Warren G (1994) Curr. Opin. Cell Biol. 4: 220-233
Schekman R (1992) Curr. Opin. Cell Biol. 4: 587--592
Stenmark H, Parton RG, Steele-Mortimer O, Lütcke A, Gruenberg J and Zerial M (1994) EMBO J. 13: 1287-1296
Tooze J and Hollinsdhead M (1991) J. Cell Biol. 115: 635-653
van der Sluijs P, Hull M, Zahraoui M, Tavitian A, Goud B and Mellman I (1991) Proc. Natl. Acad. Sci. USA 88: 6313-6317
van der Sluijs P, Hull M, Huber LA, Webster P, Goud B and Mellman I (1992) Cell 70: 729-740
Verheyden K, Fransen M, van Veldhoven PP and Mannaerts GP (1992) Biochim. Biophys. Acta 1109: 48-54
Waterham HR, Titorenko VI, Swaving GJ, Harder W and Veenhuis M (1993) EMBO J. 12: 4785-4794
Watkins JD, Hermanowski AL and Balch WE (1993) J. Biol. Chem. 268: 5182-5192
Wilson KL and Newport J (1988) J. Cell Biol. 107: 57-68
Woodman PG and Warren G (1988) Eur. J. Biochem. 173: 101-108

Change in Membrane Fluidity after pinocytosis of a new Cytokine IK on Resting and Activated T cells

[1]Raffaele Pereno, [1]Patricia KRIEF and Zohar MISHAL*

*C.N.R.S UPS 47
Laboratoire de Cytométrie
7 rue Guy Moquet
94800 Villejuif

INTRODUCTION

Pinocytosis is a mechanism by which the majority of eukaryotic cells internalize external fluid. However, endocytosis has come to be used synonymously with pinocytosis. When considering pinocytosis, it is important to distinguish between two types of uptake. One involves specific sites on the cell surface, as morphologically or biochemically distinguished, where surface binding is important and the uptake may depend on special mechanism. The other involves a non selective uptake of solute molecules that take place by surface vesiculation and engulfment of medium and does not seem to use specific binding sites (Van Deurs et al., 1990).
The membrane fluidity in human T lymphocytes is a well known process. When T lymphocytes are activated in response to specific antigens or mitogenic lectins, we observe increase in cells size followed by the enhancement of the synthesis of membrane phospholipids and by the change in their fatty acid composition and this causes a deacylation-reacylation reaction (Schinitzky M., 1984). The mitogenic lectin phytohemagglutinin (PHA) induces within 48h, lymphocyte activation and increases significantly the membrane fluidity (Anel et al., 1990).
MHC class II play a fundamental role as restriction molecules for antigen presentation to the TCR of CD4+ cells. Surface expression of MHC class II products is restricted to APCs including activated T cell (Silica P., 1994). The role of HLA class II antigen in the control of the immune response is determined not only by the genetic polymorphism of these molecules but also by their density on the cells surface (Janeway et al., 1984); indeed different groups (Lechler et al., 1985; Matis et al., 1983) have shown that the extent of T cell activation can be

[1]INSERM U268 Hôpital Paul Brousse, 1 avenue Paul Vaillant Couturier, 94807 Villejuif, France.

NATO ASI Series, Vol. H 95
Flow and Image Cytometry
Edited by Alain Jacquemin-Sablon

predicted by a function of the product of antigen concentration and the number of HLA class II molecules. In these experiments we show the effects of a new cytokine (IK) on human peripheral blood T lymphocytes undergoing blastic transformation. This cytokine down-regulates the expression of HLA class II antigens (Krief et al., 1994).

We have studied the endocytosis of IK on B cells by LSCM to determine the mechanism by which IK inhibit HLA class II expression, and we have shown the influence of IK on T cells labelled with a lipophilic fluorescent probe TMA-DPH:1-(4 Trimethylammonium)-6-phenyl -1,3,5-hexatriene that enables the determination of membrane fluidity by measurements of fluorescence polarization on a single cell level by flow cytometry.

MATERIALS AND METHODS

Immunofluorescence

Prior to labelling, cultured BLS1 cells were pre incubated for 30 min at 37°C in a serum-free RPMI 1640 medium to deplete them of any endogenous bovine transferrin that could have been internalized from the culture medium (Grant, J., 1979). Cells were then incubated for 60 min at 37°C in serum-free RPMI 1640 medium containing 50µg/ml of transferrin-FITC and recombinant protein IK labelled with biotin. After washing twice with phosphate buffered saline (PBS), cells were fixed in 0.25% paraformaldehyde in PBS for 1h at 4°C, permeability of the cell membranes was obtained with 0.2% Tween 20 in PBS for 15 min at 37°C. Cells were washed with 0.1% azide, 0.2% FCS. in PBS. Cells were incubated for 20 min at 4°C with streptavidin-RED^{670tm} (1/100) Gibco and were plated on glass cover slip pre-treated with polylysine (Mazia, D., 1975) and viewed under a LSCM.

Fluorescence laser scanning confocal microscopy

Labelled specimens were scanned with an ACAS 570 Interactive Laser Cytometer (Meridian Instruments, Inc., Okemos, MI, USA), equipped with confocal optics. The system consisted of a 5W argon ion laser tuned to 488 nm, an Olympus IMT-2 inverted microscope with a 100x oil immersion objective (N.A.1.3) Z-axis control, an XY scanning stage, and a variable pinhole aperture, all under 80486 computer co-ordination. The ACAS 570 was set up in fluorescence mode. Recombinant-IK-streptavidin-RED 670nm or transferrin-FITC were excited at 488nm. The computer converts fluorescence intensity into colour as output image. The excitation was done at 20 mW power and the filter constitution for emission detection was then conventionally used for fluorescein 520nm BP and streptavidin-RED 670nm BP. To have a good resolution, pinhole and photo multiplier voltage were set at 40 and 25% respectively.

Fluorescence polarization

This technique, based on the well-established theory of Shinitzky (Shinitzky et al.), was used with the probe TMA-DPH embedded in hydrophobic regions of cell membranes. The cells were washed 3 times in PBS, labelled with 2×10^{-6} M DPH in PBS at 20°C for 30 min, and steady-state polarization (P) measurements were performed with a FACS 440 (Gorvel et al 1984)

RESULTS AND DISCUSSION

The majority of mammalian cell types never display HLA class II molecules expression that is restricted primarily to immunocompetent (B cells and activated T lymphocytes) , macrophages and other antigen presenting cells. In most instances these cells require IFN-gamma to induce HLA class II expression (for review see Benoist and Mathis, 1990).

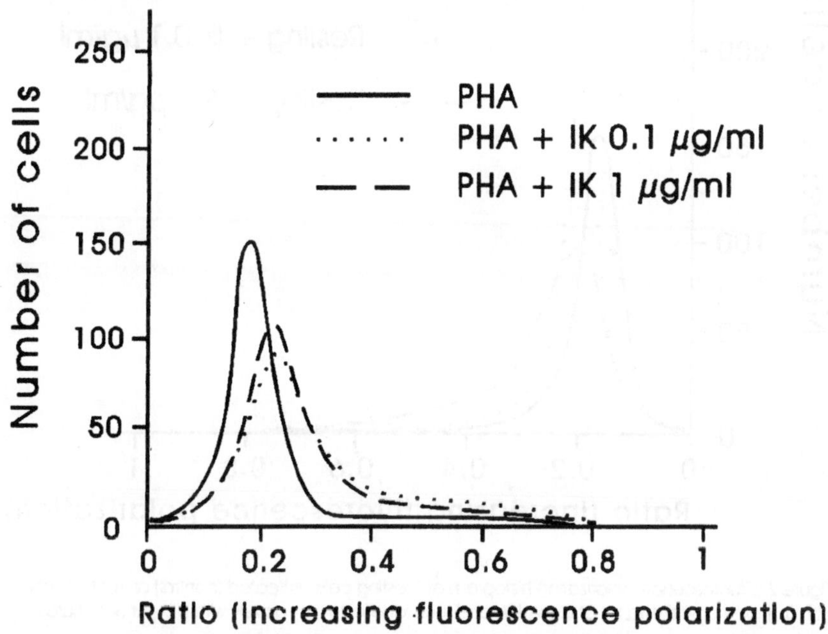

Figure 1 : Fluorescence polarization histogram of activated T cells with 2μg/ml of PHA; untreated (control) or treated either with 0.1μg/ml or with 1μg/ml of IK for 48h and then labelled immediatelly with DPH for 30 minutes.

To elucidate the potential role of the cytokine IK, down-regulating HLA class II, we have studied the changes in fluorescence polarization both on resting and on activated T cells by flow cytometry.

In these experiments we have used the cytokine IK to define its ability to modify membrane fluidity on resting and on activated T cells.

After 48h of incubation with different concentrations of inhibitor IK, we observed a significant increase of fluorescence polarization on resting and on activated T lymphocytes (fig.1). The DPH fluorescence polarization technique shows the different sensibility on activated T cells by PHA at 2μg/ml.

When IK was used at 1μg/ml on resting T cells, IK decreases the values of membrane fluidity (fig. 2). A lower dose of IK (0.1μg/ml) was necessary on activated T cells to have an increase of polarization (fig.1). On activated T cells the number of receptors would be increased and consequently the kinetic binding cytokine IK.

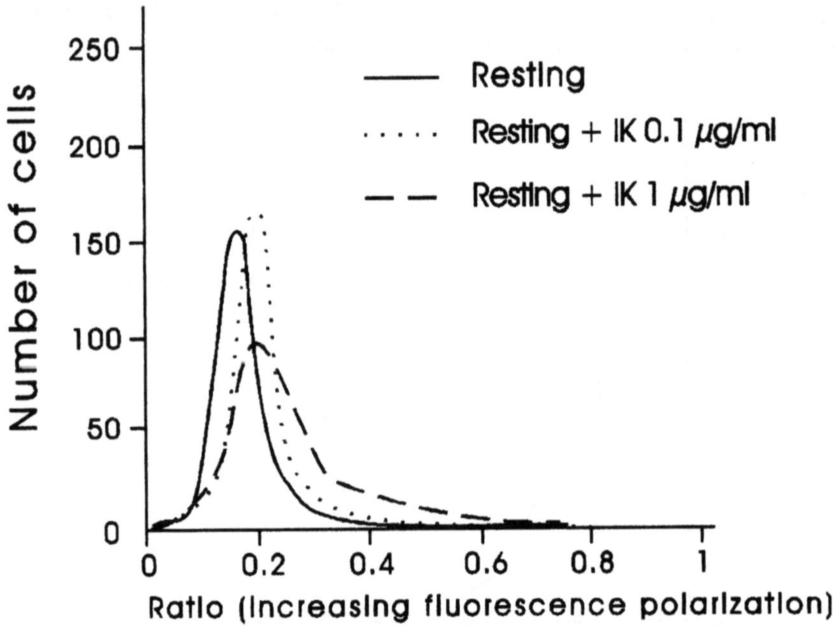

Figure 2 : Fluorescence polarization histogram of T resting cells untreated (control) or treated either with 0.1μg/ml or with 1μg/ml of IK for 48h and then labelled immediately with DPA for 30minutes.

The role of increased viscosity is classically correlated with two membrane processes 1) increase of the ratio cholesterol/phospholipids and 2) decrease of the proportion of unsaturated fatty acids in membrane phospholipids. Recent evidence shows that, on the fibroblasts FR3T3 cell line and on the mTT4 cell line, a decrease of fluidity of the cell membrane, is induced by terpene, and, biochemical analysis shows that the ratio cholesterol/phospholipids is increased by terpene (Rodes et al., 1995). Natural terpene is also capable of inhibiting proliferation of transformed mammalian cells.

Since IK may induce the increase of the ratio cholesterol/phospholipids, as it has been described for the natural terpene. This may be used to produce a decrease in membrane fluidity and therefore inhibiting T cell activation. The histogram (fig.2) shows the effect of the recombinant protein IK on resting T cells.

The localization of binding and endocytosis of the recombinant protein IK stained with streptavidinRED on Bare Lymphocyte Syndrome cells (BLS1) was examined by LSCM (fig.3)

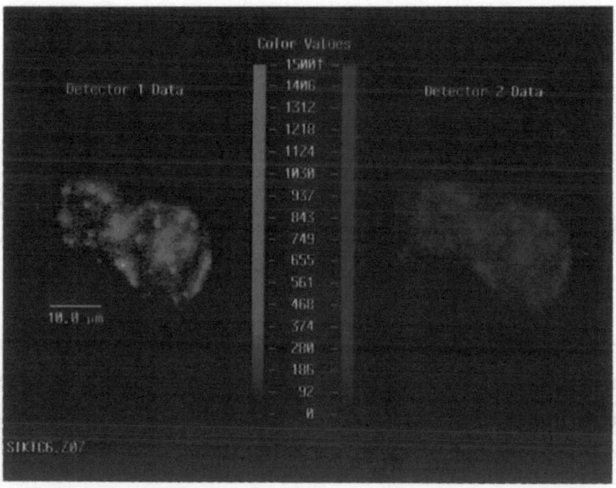

Figure 3 Co-distribution of the recombinant protein IK (Streptavidin-RED, right panel) with transferrin (Fluorescein, left panel).

These cells exhibited a punctate labelling pattern that appeared to be localised at the cell surface (streptavidinRED, right panel). The localization of the recombinant IK at the intracellular compartment, clearly reflects the endocytosis of IK by BLS1 cells that is supposed to occur through interaction of the protein with specific IK-receptors at the cell surface, this hypothesis is supported by the similar pattern observed with transferrin. Simultaneously the transferrin fluorescein conjugate was used as control. In the BLS1, transferrin enters and distributes in the cytoplasm with kinetics similar to that found with IK. The codistribution of the recombinant IK (streptavidinRED, right panel) with the transferrin fluorescein conjugate (fluorescein, left panel) indicates a co-localization.

REFERENCES

Anel A., Naval J., Gonzales B., Torros JM., Mishal Z., Uriel J. and Pineiro A. (1990) Fatty acid metabolism in human lymphocytes. I.Time-course changes in fatty acid composition and membrane fluidity during blastic transformation of periferal blood lymphocytes. Biochim. Biophys. Acta 1044 pp 323-331

Benoist, C. & Mathis, D. (1990) Regulation of Major Histocompatibility complex class II genes: X, Y and other letters of alphabet. Annu. Rev. Immunol., 8, 681-715.

Gorvel JP., Mawas C., Maroux S. and Mishal Z. (1984) Flow cytometry is a new method for the charaterization of intestinal plasma membrane. Biochem. J., 221, pp 453-457.

Grant, J. and Scrutton M. (1978) Amines inhibit the clustering of α_2-macroglobulin and EGF on the fibroblast cell surface. Nature Vol.277 22 February 661-663

Janeway,C. A., Bottomly JR., Babic K., Conrad J., Conzen P.J., Jones S., Kaye B., Katz, J.,McVay M., Murphy L., D.B. and Tite J.P. (1984) Quantitative variation in Ia-antigen expression plays a central role in immunoregulation . Immunol. Today, 5, 99-105.

Krief, P. Augery-Bourget Y., Plaisance S., Merck M., Assier E., Tanchou V., Billard M., Boucheix C., Jasmin C. & Azzarone B. (1994) A new cytokine (IK) down-regulating HLA class II : Monoclonal antibodies, cloning and chromosome localization. Oncogene, 9, 3449-3456.

Lechler R.I.,Norcross M.A., and Germain R.N. (1985) Qualitative and quantitative studies of antigens-presenting cell function by using I-A-expressing L cells. The Journal of Immunology. Vol 135 pp 2914-2922.

Matis L.A., Glimcher L.H., Paul W.E. and Schwartz R.H. (1983) Magnitude of response of Histocompatibility-restricted T-cell clones is a function of the product of the concentrations of antigen Ia molecules. Proc.Natl.Acad.Sci. USA Vol 80, pp 6019-6023.

Mazia,D., Schatten G. and Sale W. (1975) Adhesion of cells to surfaces coated with polylysine. Journal of Cell Biology Vol 66, 198-200

Rodes JF., Bonnenfant J., Trémolières A., and Brown S. (1995) Modulation of membrane fluidity and lipidic metabolism in transformed rat fibroblasts induced by the sesquiterpenic hormone farnesylacetone. Cytometry 19: 217-225.

Shinitzky, M.(1984) The lipid fluidity of cell membranes. CRC Press Inc. (ed) Physiology of Membrane Fluidity Vol I, Boca Raton Florida, pp 1-57

Shinitzky, M. and Y. Barenholz, (1978) Fluidity parameters of lipid regions determined by fluorescence polarization.Biochim. Biophys. Acta 515, 367-396.

Sillacci, P., A. Mottet, Steimle V., W. Reith, and B. Mach (1994) Developmental extinction of Major Histocompatibility Complex class II gene expression in plasmocytes is mediated by silencing of the transactivator gene CIITA. J. Exp. Med., Vol.180 1329-1336.

Van Deurs, B., O.W. Petersen, S. Olsnes, K.Sandvig, (1990) The ways of endocytosis. Int. Rev. Cytol., Vol. 117 131-177.

Acknowledgments: We thank Arlette and Valerie Vervisch for valuable advice and Domenique Laird for preparation of the manuscript.

PART II

Cell Proliferation and Gene Expression

Cell Cycle and Cell Proliferation Markers

Harry A. Crissman and Anthony J. Nastasi
Life Sciences Division
Los Alamos National Laboratory
LS-1, MS M888
Los Alamos, NM 87545

Introduction

Current concepts in cell biology suggest that there are checkpoints in the various phases of the cell cycle at which cells are monitored for damage to their DNA. Studies in our laboratory on the effect of radiation on cells in culture have shown delay of progression in both the G1 and G2 phases of the cell cycle of normal, non-transformed cells. It has been proposed that the delay period in these phases allows time for cells to repair DNA damage before proceeding to the next phase of the cell cycle. Thus cells in G1 would be required to repair potential DNA damage prior to entering S phase in order to alleviate the possibility of replicating an abnormal and potentially malignant genome. Studies on the regulations at the proposed checkpoints are important for elucidating cell cycle control mechanisms. Such investigations require techniques that can continuously monitor the distribution of cell throughout the cell cycle and access the proliferation potential of cells in different phases of the cell cycle.

Quantitative fluorescent staining and flow cytometric (FCM) analysis of cellular DNA content remains one of the most rapid and reliable approaches for determining cell cycle frequency distributions of cell populations. The precision in these studies relies upon the specificity of the staining methods for tagging DNA in cells and the efficiency of flow instruments for quantitative fluorescence analysis. Early studies, comparing computer-fit analysis of DNA content histograms to data obtained by conventional tritiated-thymidine labeling and autoradiography confirmed the accuracy of the FCM technique for cell cycle analysis (Dean and Jett, 1974).

Useful as these methods may be, the major limitation of DNA content-cell cycle frequency analysis is that it provides no information on the cycling capacity of a population and the rate at which cells traverse the cell cycle. The cell cycle frequency histogram provides a static image of the cell population showing only the

NATO ASI Series, Vol. H 95
Flow and Image Cytometry
Edited by Alain Jacquemin-Sablon

position of all cells at a given moment, but the proportions of cycling and noncycling cells within the various phases are not apparent. Sequential sampling of the population during an experiment does alleviate this problem to some extent.

The development of multiparameter flow cytometry has provided for analysis of additional physiological parameters, including DNA metabolism and other mechanism involved in the regulation and control of cell proliferation. Studies utilizing, bromodeoxyuridine (BrdUrd) incorporation have lead to the refinement of analysis of cell cycle progression. In addition, the use of monoclonal antibodies in FCM immunofluorescence studies has given rise to techniques for examining and quantitating other proliferation markers at different phases of the cell cycle. The potential of such studies is beginning to be realized not only in the area of cellular but molecular biology as well.

Methods for Cellular Bromodeoxyuridine Assay

Bromodeoxyuridine (BrdUrd)-labeling of cells in S phase or cells that have traversed S phase at some time in their growth, combined with subsequent, simultaneous FCM determination of BrdUrd content and cell cycle position, provides a powerful approach for measuring cell kinetics (Dolbeare et al., 1983, 1990). In most instances BrdUrd labeling is achieved using monoclonal antibodies in a procedure requiring that the DNA be denatured to some extent to make the BrdUrd accessible to the antibody (Gratzner, 1982)). Modification of the monoclonal antibody method vary primarily in the procedures by which the DNA is denatured. Acid-denaturation of DNA using HCl is probably most commonly used. Heat denaturation may perhaps provide more binding site for the antibody and thus be a more sensitive approach; however, the technique often produces significant loss of some cell types. Finally an enzymatic method, incorporating both endonucleases and exonucleases for exposing single-stranded BrdUrd-substituted DNA (Dolbeare and Gray, 1988), avoids the more severe acid and heat denaturation methods, but this method is more expensive and usually reserved for experiments in which acid or heat treatment might destroy other markers of interest, such as cell surface antigens.

Another method has just been developed recently by Li et al.(1995) that identifies DNA proliferating cells by selective DNA strand breaks induced by photolysis (SBIP) at sites containing incorporated BrdUrd or IdUrd. The DNA strand breaks are labeled with biotin- or digoxygenin-conjugated deoxynucleotides in a reaction catalyzed by exogenous terminal deoxynucleotidyl transferase. Fluorescein tagged avidin or digoxygenin antibody are used in the second step of the reaction. Alternatively fluorescently labeled deoxynucleotides can be used in the first step of the reaction. Cells are then counterstained with propidium iodide for analysis of total DNA content. This method is quite gentle compared to the acid or heat denaturation method and two color, red/green, fluorescence analysis can be performed using a single laser tuned to 488 nm, in a manner similar to that used for analysis of cells following anti-BrdUrd labeling.

Most other FCM procedures utilize a cytochemical approach for BrdUrd analysis that takes advantage of quantitative BrdUrd fluorescence quenching of the DNA-binding fluorochromes, Hoechst 33258 or 33342, when bound to A-BrdUrd regions in double stranded DNA (Latt, 1973). Often another DNA fluorochrome, that is not significantly affected by BrdUrd, usually ethidium bromide (EB), is used in combination with HO to obtain the cell cycle position of BrdUrd-labeled cells. Analysis can often be performed using a single uv excitation source. A number of investigators (Bohmer, R. M., 1979, 1981; Bohmer and Ellwart, 1981; Kubbies and Rabinovitch, 1983, Ellwart and Dohmer, 1985, Poot et al. (1990) have used this approach.

We have modified the BrdUrd/Hoechst quenching technique in the development of a FCM method for the analysis of populations of cells following pulse or continuous BrdUrd labeling (Crissman and Steinkamp, 1987; Crissman and Steinkamp, 1990). The procedure involves staining with both Hoechst 33342 (HO) and G-C binding mithramycin (MI) (Crissman and Tobey, 1974; Crissman et al., 1979), a dye whose stoichiometry for DNA content is not appreciably affected by BrdUrd. Using dual-laser excitation, and performing differential fluorescence analysis on the two fluorochromes, Hoechst quenching is quantitated, and thus, indirectly the cellular BrdUrd content. The technique requires only mild, one-step staining which mininizes loss of cells and other important cellular markers such as RNA, and proteins, including cellular antigens.

Comparison of Pulse and Continuous BrdUrd Labeling Experiments

In pulse labeling studies, the time duration of the BrdUrd exposure period is adjusted to only label cells in S phase. Excessive exposures period will allow labeled cells to enter G2. Following labeling cells are rinsed, cultured in BrdUrd-free medium and harvested at predetermined time intervals thereafter. FCM analysis yields information on the cycle progression capacity of cells that were in S phase during the labeling period. Methods have been devised to use such data to calculate the rate of DNA synthesis (Begg et al.,1985).

Continuous BrdUrd labeling studies are designed to, potentially, determine the frequency and cell cycle position of cells capable of synthesizing DNA over the labeling period. In these studies, cultured cells are continuously exposed to BrdUrd and harvested at desired time intervals. Analysis and calculations of the percentage of labeled and unlabeled cells in different phases of the cell cycle at different time intervals reveals the progression of cells in and out of the different phases.

Considerations for Labeling Cells with BrdUrd

Stock solutions of BrdUrd (5-bromo- 2 deoxyuridine) are freshly prepared in distilled water (1.0 mg/ml) and sterilized by filtration. The BrdUrd solution should be refrigerated in vials wrapped in foil to protect it from light, since the compound is light degradable.

For pulse labeling studies, cells are routinely labeled in culture with 30 μM BrdUrd for 1h at 37°C. Cultures should not be exposed to strong light since cells incorporating BrdUrd into their DNA can suffer chromatin damage, fail to incorporate sufficient quantities of BrdUrd or even lose viability. The concentration of BrdUrd and the length of the labeling period can vary depending on the cell cycle time and the rate of DNA synthesis. Rapidly growing may require only a 30-min pulse, but more slowly growing cells may require pulses of 1-2 h for incorporating detectable quantities of BrdUrd. Therefore, the pulse duration as well as the BrdUrd concentration must be determined empirically. Excessively high concentrations of BrdUrd are toxic to cells, and growth studies comparing of BrdUrd-treated and untreated cell populations are necessary to determine optimal conditions for BrdUrd labeling. These same criteria apply for continuous BrdUrd labeling and, in particular, the BrdUrd concentration must be reduced compared to pulse studies, since the cellular BrdUrd content is continuously increasing during the course of the study. The sampling intervals after continuous labeling can be gauged from the cell cycle doubling time and the particular cycle intevals of interest.

Cell Preparation and Staining for Differential Fluorescence Analysis

Cells are harvested from culture by centrifugation and fixed in ethanol at a concentration of about 70%. Prior to staining, fixed cell samples are centrifuged and the ethanol fixative removed by aspiration. The cells are resuspended in a dye solution containing HO and MI at concentrations of 0.5 μg/ml and 5.0 μg/ml, respectively, in PBS supplimented with 5.0 mM $MgCl_2$. Cell samples can be analyzed in the stain solution from 30 min to 5 h following staining at room temperature.

Preliminary experiments with untreated cells are performed to determine the lowest concentrations of HO and MI that can be combined to obtain DNA content histograms of good quality with low coefficient of variation (CV) values. Under these conditions the dye-dye interactions will be minimal and the subtraction of the HO fluorescent signal from the MI signal in unlabeled cells, on a cell by cell basis, will be as close to the ideal value of zero as possible.

Instrumentation Setup for Differential Fluorescence Analysis

For studies in our laboratory a multi-laser , FCM system (Steinkamp et al, 1982) is used, with one laser operating in the uv (333.6-363.8 nm) and an other tuned to 457.9 nm. The laser beams are separated by 250 μm to provide sequential excitation and analysis of each fluorochrome (ie, HO and MI respectively). A 500 nm long pass dichroic filter is used and the HO fluorescence is measured over the 400-500 nm range, and MI fluorescence above 500 nm. The electronic signal gains are adjusted so that the G1 peaks of unlabled cells are initially at the same channel number in both the HO- and MI-DNA content histograms. This initial adjustment of the G1 peak positions of unlabeled cells in both histograms sets the zero value for the subsequent subtraction process across the cell cycle. The fluorescence signals are then subtracted electronically (ie, MI minus HO) on a cell by cell basis using a differential fluorescence analysis method described by Steinkamp and Stewart, (1986). These difference measurements reflect quantitatively the quenching of HO fluorescence, which is directly proportional to cellular BrdUrd content.

In continuous labeling studies, postmitotic, BrdUrd-labeled cells may eventually appear after the initial labeling period and a storage oscilloscope (X-Y dotplot) is used to visualize the HO vs MI displays so that during data acquisition, the unlabeled G1 cells can easily be detected for alignment of the HO- and MI-G1 peaks. When all cells become labeled, then a non-BrdUrd labeled sample is used for G1 peak alignments, prior to running the labeled sample.

Differential Fluorescence Analysis of BrdUrd in Irradiated Populations

Data shown in Figure1 were obtained from analysis of human skin fibroblast (HSF) cells initially exposed to 400 R gamma irradiation and then continuously labeled with 10 μM (control) or 20 μM (irradiated culture) BrdUrd for 20 h. The DNA content, based on fluorescence from MI is on the X axis, and BrdUrd content, as derived from the MI minus the HO difference measurements is represented on the Y axis.

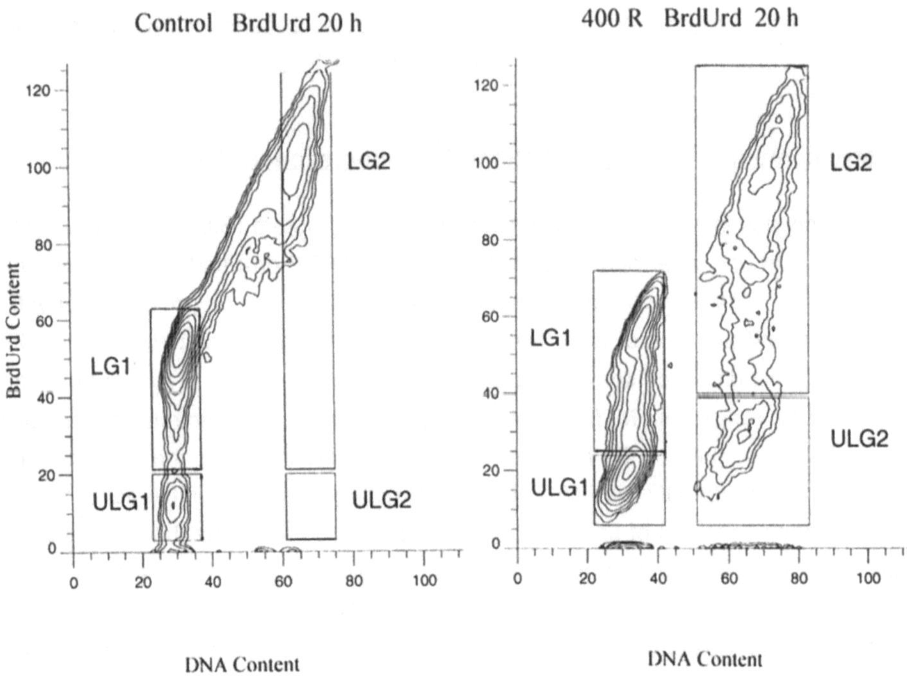

Figure 1. Bivariate contour plots of normal, human skin fibroblast (HSF) cells analyzed for DNA and BrdUrd contents following 400 R gamma irradiation and subsequent continuous BrdUrd labeling for 20 h. Unlabeled (UL) and labeled (L) subpopulations in G0/G1 and G2/M phases are delineated.

Compared to the untreated control population (Fig 1A), the irradiated cells are nearly devoid of S phase cells and are arrested in both G1 and G2 phase. Both of these compartments contain both unlabeled (UL) and labeled (L) subpopulations. Most of the control cells, with the exception of 5 % in G1, are labeled and continuing to cycle. Computer analysis of the DNA histogram for the control population at T=0 h (just prior to irradiation and BrdUrd labeling) showed approximately 60% of the cells were in GI and 20% in G2, representing the initial subpopulations of UL cells. Calculations of the data for the irradiated population indicate that 7% of the 20% (or 35% of the total UL G2 cells) initially in the G2 compartment were unable to divide. Similar analysis of HSF cells following 800 R irradiation showed that virtually all of the G2 cells originally on G2 were unable to divide by at least 20 h, indicative of a dose dependent effect of irradiation on G2 cells. However, it was noted that some label G2 cells were able to divide but others did not. Detailed analysis the freqency of labeled and unlabeled cells in the different compartment will provide detailed information on cycle progression through all phases of the cell cycle following gamma irradiation and should yield useful data for gauging the checkpoint controls imposed in the various phases

We have used these same types of analyses to detect a subpopulation of slowly traversing G1 phase cells following synchronization of human skin fibroblast (HSF) cells (Tobey et al., 1989) and to monitor the progression of HSF cells following treatment with the protein kinase inhibitor, staurosporine (Crissman et al., 1991).

The simplicity, sensitivity and accuracy of BrdUrd analysis by differential fluorescence analysis makes the method quite applicable for basic and clinical studies. Such repetitive studies require simple and rapid sample preparation, free of multiple manipulations that tend to induce cell loss and sample to sample variablility. The preparative and analytical procedure described are ideal: one-step ethanol fixation, preventing any cell loss; one-step staining with a stain cocktail and differential fluorescence analysis in the stain solution. Although most commercial FCM systems lack on-line capabilities for differential fluorescence analysis, it is possible to aquire the data in list mode and perform the subtraction of the MI and HO fluorescence signals on a cell-by-cell basis using an appropriate computer program (Habbersett et al.,1990). The only significant disadvantage is that this method requires two lasers, one with uv capabilities for exciting HO for analysis of fluorescence quenching by BrdUrd. Ideally dyes with properties similar to HO, but with excitation in the visible range, will soon become available.

Cell Cycle Phase-Specific Cyclin Assay

The cyclins represent a family of proteins that are expressed at different regions in the cell cycle. For example, cyclins D and E are expressed primarily in G1 phase, cyclin A during S and cyclin B in G2. This is a somewhat generalized scheme and there are other cyclins of interest as well. The development of antibodies to the cyclins provides the potential for measuring their expression along with DNA content by flow cytometry. Since the cyclins are cell cycle phase-dependent, a measure of their expression provides not only an independent marker for cycle position, but in addition yields information regarding the progression of cells across a particular phase.

In eukaryotic cells, mitosis is initiated after the activation of a protein kinase often referred to as maturation promoting factor (Murray and Kirschner, 1989). This protein kinase is comprised of a catalytic subunit (cdc2), a regulatory subunit (cyclin B) and a low molecular weight subunit p13-Suc 1 (Gautier et al., 1990). Following mitosis, cyclin B is degraded sometime near anaphase. If cyclin B is not degraded mitotic arrest can occur, and if cyclin B degradation is extensively delayed, mitotic abnormalities become apparent.

Sherwood et al. (1994) used FCM to for cyclin B/DNA content analysis on Hela cells treated with the cell cycle perturbing agent, aphidicolin. Cells treated with 0.2 μg/ml of aphidicolin were shown to undergo unbalanced growth such that cyclin B accumulation did not occur for 24 h compared to 13 h in the control cells. Gong et al. (1993) used similar FCM analysis to demonstrate two different DNA ploidy levels based on cyclin B measurement. In those studies, polyploid G1 cells could be distinguished from diploid G2/M cells in cultures of MOLT-4 populations treated with 0.1 μM staurosporine for varying periods of time. The polyploid G1 cells were identified by their low levels of cyclin B compare to the diploid G2/M cells with the same DNA content.

Immunofluorescent techniques for labeling cells with anti-cyclin B antibodies involve formaldehyde or ethanol fixation followed by detergent permeablization and indirect antibody labeling. A secondary fluorecein-conjugated antibody is used and cells are counter stained with propidium iodide for correlated cyclin B/DNA content determinations by FCM. The FCM method has several advantages including the fact that cells need not be synchronized for routine studies and also cells with varying levels of cyclin B can be detected and quantitated.

Cyclin B Analysis in HeLa Cells

Figure 2 shows an example of cyclin B/DNA content analysis of HeLa cells treated only with the secondary FITC-conjugated antibody (Control), and an exponentially growing (Exponential) culture and a nocodazole-treated culture. Cells were treated for 12 h with 0.1 µg/ml of nocodazole to accumulate cells in the G2/M phase of the cell cycle to better demonstrate the specificity of the technique. The exponentially growing cells show some cyclin B labeling in G2 phase but levels of cyclin B are considerably increased in the nocodazole-treated cultures where cell division is delayed. The analysis of cyclin B and other cyclins will undoubtly become a routine FCM procedure and the method definitely adds an additional FCM parameter that is of considerable importance in cell cycle-related studies.

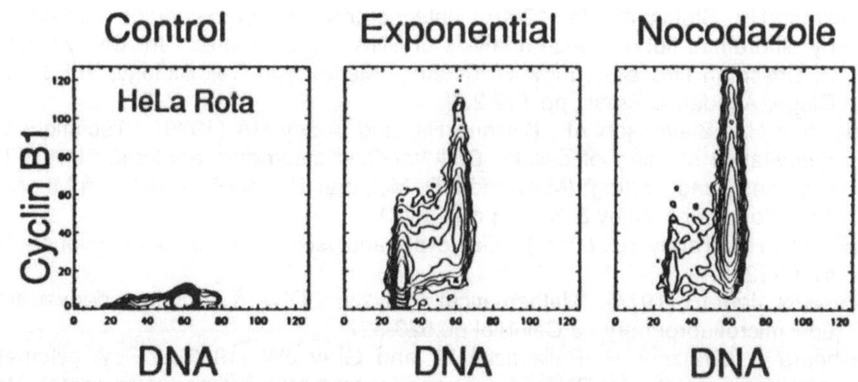

Figure 2. Bivariate contour plots of HeLa Rota cell populations analyzed for DNA and cyclin B1 contents. One population (Control) was not treated with mouse anti-cyclin B1 antibody but only with the fluorescein-conjugated anti-mouse antibody. The other distributions show analysis of an asynchnous population (Exponential) and a population that was treated with 0.1 µg/ml nocodazole for 12 h (Nocodazole).

Acknowledgments

This work was supported by the Los Alamos National Flow Cytometry and Sorting Resource (NIH RR013151) and National Institutes of Health Grants RR07855 and R01 RR06758 and the U.S. Department of Energy.

References

Begg AC, McNally NJ, Shrieve DC, Karcher HA (1985). A method to measure the duration of DNA synthesis and the potential doubling time from a single sample Cytometry 6:620-626.

Bohmer RM (1979). Flow cytometric cell-cycle analysis using the quenching of 33258 Hoechst fluorescence by bromodeoxyuridine incorporation. Cell Tissue Kinet. 12:101-110.

Bohmer RM, Ellwart J (1981) Combination of BUdR-quenched Hoechst fluorescence with DNA-specific ethidium bromide fluorescence for cell cycle analysis with a two-parameter flow cytometer. Cell Tissue Kinet. 14: 653-658.

Crissman HA, Gadbois DM, Tobey RA, Bradbury EM. (1991) Transformed mammalian cells are deficient in kinase-mediated control of progression through the G1 phase of the cell cycle. Proc. Natl Acad. Sci, USA 88-7580-7584.

Crissman HA, Steinkamp JA (1987) A new method for rapid and sensitive detection of bromodeoxyuridine in DNA replicating cells. Exp. Cell Res. 173: 256-261.

Crissman HA, Steinkamp JA (1990). Detection of bromodeoxyuridine-labeled cells by differential fluorescence analysis of DNA fluorochromes. In: Darzynkiewicz Z, Crissman HA (eds), Flow Cytometry. Methods in Cell Biology, Vol 33, San Diego: Academic Press, pp 199-206.

Crissman HA, Stevenson AP, Kissane RJ, and Tobey RA (1979). Techniques for quantitative staining of cellular DNA for flow cytometric analysis. In: "Flow Cytometry and Sorting" (Melamed MR, Mullaney PF, and Mendelsohn ML, eds), New York: John Wiley & Sons, pp 243-261.

Crissman HA, Tobey RA (1974). Cell cycle analysis in 20 minutes. Science 184: 1297-1298.

Dean PN, Jett JH (1974). Mathematical analysis of DNA distributions derived from flow microfluorometry. J Cell Biol 60:523-527.

Dolbeare F, Gratzner H, Pallavacini M, and Gray JW (1983). Flow cytometric measurement of total DNA content and incorporated bromodeoxyuridine. Proc Natl Acad Sci USA 80: 5573-5577.

Dolbeare F, Gray JW (1988). Use of restriction endonucleases and exonuclease III to expose halogenated pyridines for immunochemical staining. Cytometry 9: 631-635.

Dolbeare F, Kuo W-L, Beisker W, Vanderlaan M, Gray JW (1990). Using monoclonal antibodies in bromodeoxyuridine-DNA analysis. In: Darzynkiewicz Z, Crissman HA (eds), Flow Cytometry. Methods in Cell Biology, Vol 33, San Diego: Academic Press, pp 207-216.

Ellwart J, Dohmer P (1985). Effect of 5-fluoro-2'-deoxyuridine (FdUrd) on 5-bromo-2 deoxyuridine (BrdUrd) incorporation into DNA measured with a monoclonal BrdUrd antibody and by BrdUrd/Hoechst quenching effect. Cytometry 6: 513-520.

Gautier J, Minshull J, Lohka M, Glotzer M, Hunter T, Maller JL (1990). Cyclin is a component of maturation-promoting factor from Xenopus. Cell 60:487-494.

Gong J, Traganos F, Darzynkiewicz Z (1993). Simultaneous analysis of cell cycle kinetics at two different DNA ploidy levels based in DNA content and cyclin B measurements. Cancer Res 53:5096-5099.

Gratzner H (1982). Monoclonal antibody against 5-bromo- and 5-iododeoxyuridine: A new reagent for detection of DNA replication. Science 218: 474-475.

Habbersett R, Crissman HA, Jett JH (1990). BrdU incorporation detected by differential fluorescence and digital signal processing. Cytometry (Suppl. 4), Abstract 528 B. p88.

Kubbies M, Rabinovitch PS (1983). Flow cytometric analysis of factors which influence the BrdUrd Hoechst quenching effect in cultivated human fibroblasts and lymphocytes. Cytometry 3: 276-281.

Latt SA (1973). Microfluorometric detection of deoxyribonucleic acid replication in human metaphase chromosomes. Proc Natl Acad Sci USA 70: 3395-3399.

Li X, Traganos F, Melamed MR, Darzynkiewicz Z. Single step procedure for labeling DNA strand breaks with fluorescein- or BODIPY-conjugated deoxynucleotides. Detection of apoptosis and BrdUrd incorporation. Cytometry, in press.

Murray AW, Kirschner MW (1989). Dominoes and clocks: the union of two views of the cell cycle. Science 246:614-621.

Poot M, Hoehn H, Kubbies M, Grossman A, Chen Y, Rabinovitch PS (1990). Cell cycle analysis using continuous bromodeoxyuridine labeling and Hoechst 33258-ethidium bromide bivariate flow cytometry. In: Darzynkiewicz Z, Crissman HA (eds), Flow Cytometry. Methods in Cell Biology, Vol 33, San Diego: Academic Press, pp 185-206.

Sherwood SW, Rush DF, Kung AL, Schimke RT (1994). Cyclin B1 expression in HeLa S3 cells studied by flow cytometry. Exp Cell Res 211:275-281.

Steinkamp JA, Stewart CC (1986). Dual laser, differential fluorescence correction method for reducing cellular background autofluorescence. Cytometry 7: 566-574.

Steinkamp JA., Stewart CC, Crissman H. A. (1982). Three-color fluorescence measurements on single cells excited at three laser wavelengths. Cytometry 2, 226-231.

Tobey RA., Oishi N., Crissman HA. (1989) Synchronized human diploid fibroblasts: Progression capabilities of a subpopulation that fails to keep pace with the predominant, rapidly dividing cohort of cells. J. Cell. Physiol. 139:432-440.

Advances in Flow Cytogenetics: Progress in the Development of a High Speed Optical Chromosome Sorter Based on Photoinduced Adduct Formation Between Psoralens and Chromosomal DNA

Mary C. Roslaniec, Richard J. Reynolds, John C. Martin, Ku Taek Han[1] and L. Scott Cram
Life Sciences Division, National Flow Cytometry Resource
Los Alamos National Laboratory
Los Alamos, NM 87545
United States of America

I. INTRODUCTION

The National Flow Cytometry Resource (NFCR) at Los Alamos National Laboratory has become nationally and internationally recognized for its expertise in chromosome sorting. As a result, we receive more requests than we can accommodate from research investigators worldwide in need of assistance for procedures and/or actual sorted chromosomes. We have two flow cytometers committed full time to chromosome sorting. Under ideal conditions (often sample dependent) we can sort about two million chromosomes (0.5 µg) per eight hour day with both instruments operating in parallel.

The protracted study of the human genome and now the mouse genome continues to warrant a clear need for sorting large numbers of chromosomes. The ability to identify and isolate specific genetic sequences is pivotal to genetic mapping, the search for DNA sequence probes, construction of chromosome specific libraries, identification of a diseased gene and basic genome organization. Since the human genome is comprised of 3 billion DNA base pairs distributed over 24

[1] Ku Taek Han
Department of Obstetrics and Gynecology
St. Mary's Hospital
Catholic University Medical College
62 Youido-dong, Youngdungpo-gu
Seoul 150-010
Korea

NATO ASI Series, Vol. H 95
Flow and Image Cytometry
Edited by Alain Jacquemin-Sablon

chromosome types, identification of a specific gene or mutation in this system is a complex undertaking. Chromosome sorting can be used to narrow the source DNA to a specific chromosome. By sorting human chromosomes prior to analysis, the complexity of the study is reduced by a factor of about 24. Sorted chromosomes have proven to be invaluable to the National Laboratory Gene Library Project for the construction of small insert libraries (up to 25 kb) prepared from bacteriophages and large insert libraries (35 to several hundred kb) prepared from cosmids and YACs (Griffith, 1984; Longmire, 1993; McCormick, 1993; Van Dilla, 1984).

In response to the need for large quantity chromosome sorting, the NFCR has initiated the development of a high speed optical chromosome sorter based on the photoinactivation of DNA. Chromosomes are analyzed as in traditional flow cytometry, however, rather than relying on droplet formation, (a rate limiting factor in conventional sorting), unwanted chromosomes are photodamaged rendering them unclonable. This photodamage is induced by exposure of unwanted chromosomes to a high power laser beam after they have passed the measurement stations. Desired chromosomes are then cloned in the presence of the unwanted, unclonable DNA.

We have successfully completed the first phase of the project by demonstrating the ability to photoinactivate DNA under steady state conditions. Our ultimate goal is the construction of chromosome specific libraries. Cosmid cloning with the s-Cos-1 vector is routinely used at the Center for Human Genome Studies (CHGS) at Los Alamos for this purpose. Therefore, in conjunction with this project, we have developed an assay to demonstrate the effects of DNA photoinactivation on s-Cos-1 cosmid cloning and we report the results of our preliminary studies below.

II. CURRENT TECHNOLOGIES

Chromosome sorting is currently achieved using flow cytogenetics. Briefly, monodispersed chromosomes stained with one or two fluorescent dyes pass either one or two sequential focused laser beams tuned to excite the fluorescent stains. Optimally, each chromosome has a unique staining pattern allowing identification based on distinctive fluorescent characteristics. The flow jet breaks into droplets and those droplets containing desired chromosomes are electrically charged. All droplets pass an electric field and charged droplets are deflected to a separate collection tube thus sorting the chromosomes (Albright, 1991). Analysis of a single chromosome in a flow cytometer was first reported in 1975 (Gray, 1975; Stubblefield, 1975). Initially, a single fluorescent DNA probe was used to distinguish between chromosome types; however, by 1980, bivariate analysis of human chromosomes had been developed making it possible to resolve 21 of the 24 types of human chromosomes. Improvements in buffers, (Cram, 1990b,c) high speed sorting, (Peters, 1985) slit scan analysis, and data analysis (Lucas, 1989) are among the advances made since then. Several extensive reviews of flow cytogenetics are currently available (Cram, 1990a; Darzynkiewicz, 1990; Gray, 1989; Melamed, 1990).

While flow cytometry is the principle technique used in sorting chromosomes, drawbacks persist; separation is limited by the rate of stable droplet formation. The extent of separation is inversely proportional to analysis rate, i. e., the faster the sorting, the lower the purity (Melamed, 1990, p. 571). For example, in sorting chromosome # 4 of a hamster-human cell line (UV20 HL21-27), the purity drops from 98 to 91% when the analysis rate is increased from 5000 to 18,000 chromosomes s^{-1} (Albright, 1991). Furthermore, to obtain high purity chromosomes, the probability of having more than one particle per drop must be extremely low, hence, particles are processed at approximately one tenth the rate of droplet formation. Commercial sorters operate at relatively slow chromosome sorting rates. For example, with analysis rates of 2000 chromosomes s^{-1} and sort rates of less than 50 chromosomes s^{-1}, 60 hours of sorting time are required to obtain 1 μg of DNA. With the high speed sorter developed at Los Alamos and Lawrence Livermore National Laboratories, it is possible to achieve rates on the order of 200,000 droplets s^{-1}; hence, samples can be analyzed at rates in excess of 20,000 chromosomes s^{-1} (Peters, 1985). To achieve such high sorting rates, however, the sorter must be operating at high pressure (200 psi) resulting in sorting instability and non-selective chromosome damage. By reducing the pressure to about 120 psi, droplets can be produced at rates of 140,000 s^{-1} improving sorting stability and lessening DNA degradation (Albright, 1991).

III. NEW DEVELOPMENTS

Methods of chromosome separation *not* utilizing droplet formation would eliminate problems of droplet instability and DNA damage. Photochemistry offers a possible alternative to traditional droplet sorting. Photochemical processes tend to be *fast* (on the nanosecond time scale). Chromosome sorting based on photochemical inactivation of DNA would, therefore, be favored over *slow* droplet sorting. DNA lesions can be induced by several photochemical methods including direct far UV damage, photosensitization via oxygen dependent and independent mechanisms or both. Photochemically induced lethal lesions would be expected to inactivate the DNA. Chromosomes comprised of inactivated DNA would be unclonable thus providing a means of separating desired and undesired chromosomes.

A method of DNA inactivation under consideration is photosensitization with furocoumarins. Furocoumarins, also known as psoralens, are linear tricyclic heterocycles with a furan ring fused to a coumarin moiety. Oxygen independent photochemical induction of DNA cross-linking is the main mode of DNA inactivation. The initial interaction between DNA and psoralen is the noncovalent intercalation of the psoralen into the DNA double helix. Irradiation of the complex with near UV light induces a cycloaddition reaction between the double bond of a pyrimidine base (predominately thymine) and the olefin of either the furan or pyrone moieties in the psoralen resulting in a cyclobutane adduct. This occurs via the triplet state of the psoralen which is easily attained, e. g., the $S \Rightarrow T$ quantum yield of 8-methoxypsoralen in water is high (0.14) and its E_t is 262 kJ/mol. Upon further irradiation, furan monoadducts can crosslink by undergoing a second cycloaddition reaction involving the olefin of the pyrone moiety provided there is an adjacent pyrimidine base (Averbeck, 1989; Cimino, 1985; Song, 1979). (Kinetics for this reaction are discussed below). We propose that, once cross-linked, the DNA could not denature and would be unclonable. It may also be possible to modify the cloning process such that monoadducts are lethal as well.

Rendering DNA unclonable by light induced furocoumarin deactivation has been demonstrated in the past. In 1986, Yokobata et. al developed a plaque reduction assay based on this principle to assess the biological activity of duplex circular DNA modified by covalent adduct formation of psoralen derivatives. Bacterial cells were transfected with the replicating form of psoralen cross-linked bacteriophage DNA and plated on a confluent lawn of cells. Plaque development provided a measure of infectivity of the bacteriophage; phage DNA treated with psoralen and light demonstrated decreased plaque formation (Yokabata, 1986).

In the proposed Optical High Speed Sorter, furocoumarin intercalation into the chromosomal DNA will take place prior to sorting. The chromosomes will be analyzed in the conventional manner but no droplets will be generated. Rather, unwanted chromosomes will be irradiated with a third laser tuned to the absorption of the intercalated psoralen to induce cross-linking. As desired chromosomes pass this point, this third laser beam is optically blocked by an optical modulator (see Figure 1). The desired chromosomes are not cross-linked and will subsequently be cloned. The optical high speed sorter is based on the "zapper principle" first described at Los Alamos (Martin, 1981). More recently, this concept was used for high-speed photodamage cell selection with a frequency-doubled argon ion laser (Keij, 1995). Using optical high speed sorting, we predict sorting rates of up to 1000 s^{-1}, fifty times that possible with conventional sorters.

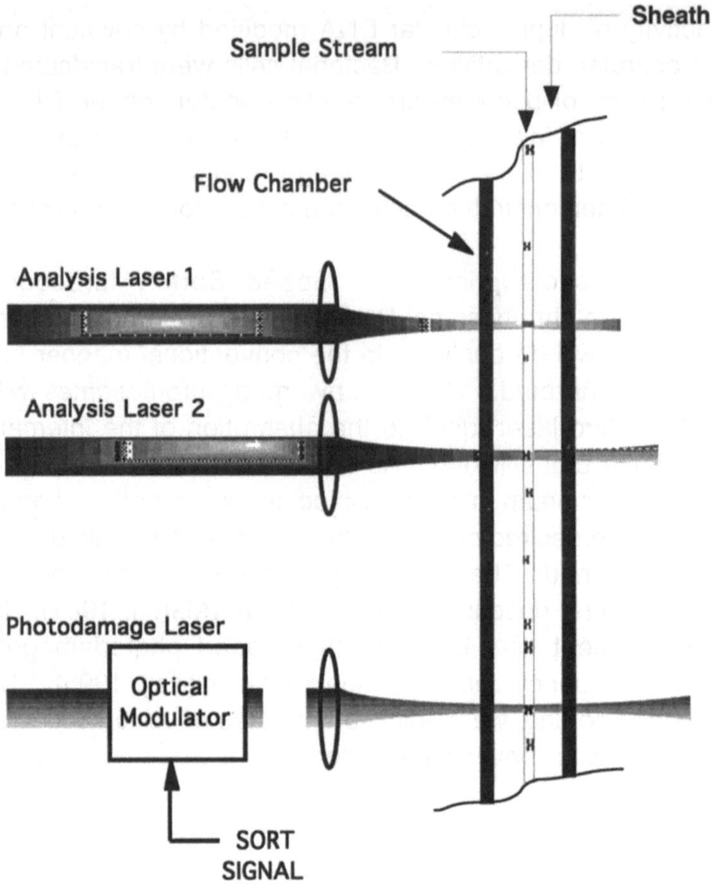

Figure 1. Schematic diagram of the proposed optical high speed sorter. The chromosome suspension is incubated with a psoralen derivative prior to sorting. Monodispersed chromosomes stained with two fluorescent dyes are analyzed by the first two lasers based on their distinctive fluorescent characteristics. A third laser controlled by an optical modulator selectively irradiates unwanted chromosomes causing cross-links between the intercalated psoralen and a pyrimidine base. Unirradiated chromosomes remain clonable.

In phase one of this project, we have established the ability to inhibit cloning by photoinactivation of DNA. The cloning assay that has been developed uses the s-Cos-1 cloning vector. This is the vector of choice of the Center for Human Genome Studies at Los Alamos for the construction of chromosome specific libraries. Using this vector, it is possible to construct high quality genomic libraries from limited (< 1 μg) quantities of DNA. Two *cos* (cohesive) sites allow for efficient cloning of non-size selected DNA while unique *Not* I, *Sac* II or *Sfi* I sites which flank the cloning site permit easy removal of the insert DNA from the vector (Evans, 1989).

The assay is performed as follows. GM130 DNA is incubated with a psoralen derivative allowing time for intercalation. The psoralen/DNA is then irradiated with long wavelength UV light (≈ 365 nm) at an estimated dose of 8.3 kJ/m^2 to induce covalent bond formation. Initial experiments were performed in a static system using a high pressure Hg arc lamp. After exposure, the DNA is digested with a restriction enzyme, dephosphorylated, ligated to s-Cos-1 vector arms and packaged into a protein coat (viral capsid) in preparation for transfection. The packaged DNA is then transfected into DH5αmcr bacteria and plated onto kanamycin containing agar plates. Since the s-Cos-1 vector arm carries a gene for kanamycin resistance, DNA which is cloned imparts kanamycin resistance to the bacteria. DNA which has been photodamaged and is therefore unclonable cannot impart such resistance. Bacterial growth (colony counts) is then observed to determine the extent of photoinactivation.

Two representative psoralen derivatives that have been assayed for their photoinactivating capabilities are trimethylpsoralen (TMP) and its water soluble analogue aminomethyltrimethylpsoralen (AMT) (see Figure 2).

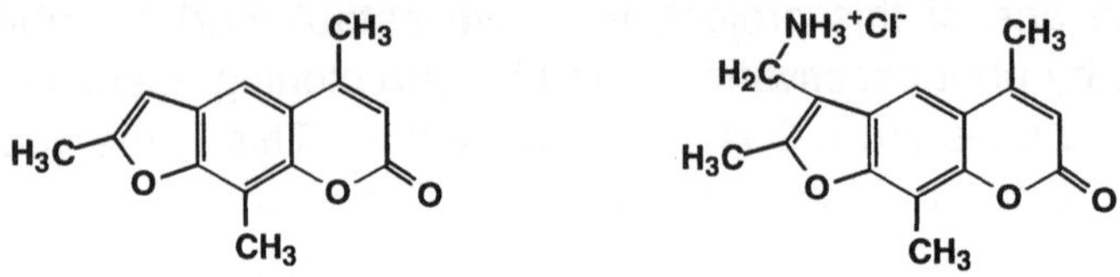

TMP = trimethylpsoralen **AMT = aminomethyltrimethylpsoralen**

Figure 2 Two psoralen derivatives, aminomethyltrimethylpsoralen (AMT) and trimethylpsoralen (TMP) that have been tested for their DNA photoinactivating capabilities.

These two psoralen derivatives have been evaluated using the cloning assay described above and the results are shown in Table 1. The cloning efficiency for DNA unexposed to psoralen or light is high ⇒ 3.3×10^5 colony forming units (CFU)/µg DNA. Addition of psoralen under 'dark' conditions has only minimal effects in the case of either TMP or AMT. Irradiating the DNA without psoralen present has a modest inhibitory effect and the cloning efficiency is reduced to 3.8×10^4 CFU/µg DNA. However, irradiation of the DNA in the presence of either psoralen has a dramatic effect on the cloning efficiency reducing it 100 and 1000 fold in the cases of AMT and TMP, respectively. Note that UV light alone can cause sufficient photodamage to reduce the cloning efficiency by a factor of 10. In the proposed optical sorter, desired chromosomes will not be exposed to the high powered inactivating laser. We include this data to demonstrate that, at least at long exposure times, a portion of the photodamage is due to UV light alone. We are considering this as an alternate form of photoinactivation. Other forms of photoinactivation that we are considering include oxygen dependent or independent photoinduced energy or electron transfer reactions.

Table 1. Effects of light induced psoralen/DNA adduct formation on cloning efficiency.

DNA treatment	-psoralen -light	+TMP -light	+AMT -light	-psoralen +light[a]	+TMP +light[a]	+AMT +light[a]
colonies[b]	1014	1309	904	117	1	16
cloning efficiencies[c]	3.3×10^5	4.2×10^5	2.9×10^5	3.8×10^4	3.2×10^2	5.2×10^3

a 8.3 kJ/m^2

b total # of colonies on five plates

c cloning efficiencies given in *colony forming units (CFU)/µg of insert DNA*

We have also demonstrated the ability to clone desired DNA in the presence of photodamaged DNA. We combined equal amounts of desired and photodamaged DNA and subjected this mixture to the cloning process. The cloning efficiency of this mixture was not significantly different to that of the desired DNA alone.

We feel that we will be able to reduce cloning efficiencies of unwanted material further by optimizing the light and psoralen doses. Furthermore, by replacing the DH5αmcr bacteria, an excision repair-*proficient* strain, with an excision repair-*deficient* strain, cloning efficiencies should decrease further. This is especially true in the case where the ratio of monoadducts to crosslinks is relatively high.

The cloning assay described above was developed in a *static* system. DNA photoinactivation in a *flow* system presents additional challenges. The major obstacle is the time dependence of the photoinactivation. It is necessary for the photoreaction to be completed within the time that the psoralen/chromosome complex crosses the light path. The estimated time the psoralen/chromosome complex will be in the path of the photoinactivating laser beam is <10 µs. We predict that this is sufficient for cross-linking. Monoaddition is fast; the rate constant for the reaction of psoralen with thymine is on the order of 10^8 M^{-1} s^{-1} (Bensasson, 1978). In the case of AMT, it has been shown that as little as two 15 ns pulses from a frequency doubled Q-switched ruby laser (347 nm) are sufficient for cross-linkage inferring that diadduct formation is also fast (Johnston, 1977). In subsequent work, Johnston et. al demonstrated that after a 15 ns pulse of the 3rd harmonic of a neodymium:YAG laser followed by a 1 μs time delay, a 10 ns pulse of a nitrogen laser was sufficient for diadduct formation; the rate constant, k_m, observed for the cross-linking step was 7.9×10^5 s^{-1}. The time delay has been attributed to the time required for the monoadduct to achieve the conformation necessary for cross-linking (Johnston, 1981). Considering these kinetics, we feel that the photoinactivation methods can be adapted to a flow system. Initial experiments in a flow system are underway.

IV. CONCLUSION

As a preliminary step for the development of the proposed optical chromosome sorter, we have successfully demonstrated the ability to inhibit cloning of undesired chromosomes under steady state conditions through photoinactivation of DNA. Furthermore, we have also demonstrated the ability to clone unexposed DNA in the presence of unwanted DNA. These DNA photoinactivation techniques are currently being tested under flow conditions.

Currently, to sort isolated chromosomes for the construction of a chromosome specific medium size insert library requires approximately one week of round-the-clock sorting. We expect that with the increased throughput rate of the optical sorter, we will be able to recover enough chromosomes in a matter of hours to construct such a library. This will benefit researchers world-wide.

ACKNOWLEDGMENTS: Support provided by the U. S. Department of Energy and NIH Grant RR01315.

REFERENCES

Albright, K. L.; Cram, L. S.; Martin, J. C. In (1991) Am. Chem. Soc. Symp. Series. Vol. 464; Separation Techniques Used to Prepare Highly Purified Chromosome Populations: Sedimentation, Centrifugation and Flow Sorting. pp 73-88

Averbeck, D. (1989) Recent Advances in Psoralen Phototoxicity Mechanism. Photochem. Photobiol. 50:859-882

Bensasson, R. V.; Land, E. J.; Salet, C. (1978) Triplet Excited State of Furocoumarins: Reaction with Nucleic Acid Bases and Amino Acids. Photochem. Photobiol. 27:273-280

Cimino, G. D.; Gamper, H. B.; Isaacs, S. T.; Hearst, J. E. (1985) Psoralens as Photoactive Probes of Nucleic Acid Structure and Function: Organic Chemistry, Photochemistry and Biochemistry. Ann. Rev. Biochem. 54:1151-1193

Cram, L. S. (1990a) Flow Cytogenetics and Chromosome Sorting. Human Cell 3:99-106

Cram, L. S.; Campbell, M.; Fawcett, J.; Deaven, L. L. (1990b) Polyamine Buffer for Bivariate Human Flow Cytogenetic Analysis and Sorting. Meth. in Cell Biol. 33:377-383

Cram, L. S.; Ray, F. A.; Bartholdi, M. F. (1990c) Univariate Analysis of Metaphase Chromosomes Using the Hypotonic Potassium Chloride-Propidium Iodide Protocal. Meth. Cell Biol. 33:369-367

Darzynkiewicz, Z.; Crissman, H., Ed. (1990) Flow Cytometry; Academic Press: San Diego ; Vol. 33

Evans, G. A.; Lewis, K.; Rothenberg, B. E. (1989) High Efficiency Vectors for Cosmid Microcloning and Genomic Analysis. Genome 79:9-20

Gray, J. W., Ed.; (1989) Flow Cytogenetics; Academic Press: San Diego

Gray, J. W.; Carrano, A. V.; Steinmitz, M. A.; Van Dilla, M. A.; Moore, D. H.; Mayall, B. H.; Mendelsohn, M. L. (1975) Chromosome Measurement and Cell Sorting by Flow Systems. Proc. Natl. Acad. Sci., U. S. A. 72:1231-1234

Griffith, J. K.; Cram, L. S.; Crawford, B. D.; Jackson, P. J.; Schilling, J.; Schimke, R. T.; Walters, R. A.; Wilder, M. E.; Jett, J. H. (1984) Construction and Analysis of DNA Sequence Libraries from Flow-Sorted Chromosomes: Practical and Theoretical Considerations. Nuc. Acids Res. 12:4019-4034

Johnston, B. H.; Johnson, M. A.; Moore, C. B.; Hearst, J. E. (1977) Psoralen-DNA Photoreaction: Controlled Production of Mono- and Diadducts with Nanosecond Ultraviolet Laser Pulses. Science 197:906-908

Johnston, B. H.; Kung, A. H.; Moore, B.; Hearst, J. E. (1981) Kinetics of Formation of Deoxyribonucleic Acid Cross-links by 4'-(Aminomethyl)-4,5',8-trimethylpsoralen. Biochemistry 20:735-738

Keij, J. F.; Groenewegen, A. C.; Dubelaar, G. B. J.; Visser, J. W. M. (1995) High-Speed Photodamage Cell Selection Using a Frequency-Doubled Argon Ion Laser. Cytometry 19:209-216

Longmire, J. L.; Brown, N. C.; Meincke, L. J.; Albright, K. L.; Fawcett, J. J.; Campbell, E. W.; Moyzis, R. K.; Hildebrand, C. E.; Evans, G. A.; Deaven, L. L. (1993) Construction and Characterization of Partial Digest DNA Libraries Made from Flow-Sorted Human Chromosome 16. GATA 10:69-76

Lucas, J.; Bartholdi, M. F. In (1989) Flow Cytogenetics. J. Gray ed. Academic Press: New York. Chromosomes Classification by Scanning Flow Cytometry. pp 173-190

Martin, J. C.; Jett, J. H. (1981) Photodamage, a Basis for Super High Speed Cell Selection. Cytometry 2:114, (abstract 91)

McCormick, M. K.; Campbell, E.; Deaven, L.; Moyzis, R. (1993) Low-Frequency Chimeric Yeast Artificial Chromosome Libraries from Flow-Sorted Human Chromosomes 16 and 21. Proc. Natl. Acad. Sci. 90:1063-1067

Melamed, M. R.; Lindmo, T.; Mendelsohn, M. L. (1990) Flow Cytometry and Sorting; 2nd ed.; Wiley-Liss: New York

Peters, D.; Branscomb, E.; Dean, P.; Merrill, T.; Pinkel, D.; Van Dilla, M. A.; Gray, J. W. (1985) The LLNL High Speed Sorter: Design Features, Operational Characteristics and Biological Utility. Cytometry 6:290-301

Song, P. S.; Tapely, K. J. (1979) Photochemistry and Photobiology of Psoralens. Photochem. Photobiol. 29:1177-1197

Stubblefield, E.; Cram, L. S.; Deaven, L. L. (1975) Flow Microfluorometric Analysis of Isolated Chinese Hamster Chromosomes. Exp. Cell Res. 94:464-468

Van Dilla, M. A.; Deaven, L. L.; Albright, K. L.; Allen, N. A.; Aubuchon, M. R.; Bartholdi, M. F.; Brown, N. C.; Campbell, E. W.; Carrano, A. V.; Clark, L. M.; Cram, L. S.; Crawford, B. D.; Fuscoe, J. C.; Gray, J. W.; Hildebrand, C. E.; Jackson, P. J.; Jett, J. H.; Longmire, J. L.; Lozes, C. R.; Luedemann, M. L.; Martin, J. C.; McNinch, J. S.; Meincke, L. J.; Mendelsohn, M. L.; Meyne, J.; Moyzis, R. K.; Munk, A. C.; Perlman, J.; Peters, D. C.; Silva, A. J.; Trask, B. J. (1984) Human Chromosome-Specific DNA Libraries: Construction and Availability. Bio/Technology 4:537-552

Yokabata, K. E.; Jordan, J. M.; Chapman, O.; Kreil, C. (1986) Development of a Plaque Reduction Assay and Application to the Study of Psoralen-Damaged DNA. Photochem. Photobiol. 43:391-401

ANALYSIS OF CELL DEATH BY FLOW CYTOMETRY

Zbigniew Darzynkiewicz and Xun Li
The Cancer Research Institute
New York Medical College
Valhalla, N.Y. 10595, USA

Introduction

Apoptosis and *necrosis* denote two distinct modes of cell death. The distinction is based on differences in morphological, biochemical and molecular changes that occur within the dying cell (reviews, Arends *et al.*, 1990; Compton *et al.*, 1992, Wyllie, 1985, 1992, Wyllie *et al.*, 1980). Apoptosis, often called *cell suicide* or *programmed cell death,* is a particular mode of death in which the affected cell, by activating a characteristic cascade of biochemical events which ultimately lead to its total physical disintegration, actively carries on the program of its demise. The early event of apoptosis is cell dehydration. Loss of intracellular water results in a change of cell shape and size: the originally round cells become elongated or convoluted in shape and diminished in size. Another characteristic change during apoptosis is chromatin condensation. The condensation starts at the nuclear periphery and is followed by nuclear fragmentation. A very distinct hyperchromicity and homogeneity characterizes the stainability of DNA in the condensed chromatin of apoptotic cells. At later stages of apoptosis, the nuclear fragments, together with the constituents of the cytoplasm (including intact organelles), are packaged into so called *apoptotic bodies*, which detach from the dying cell. When apoptosis occurs *in vivo,* the apoptotic bodies are phagocytized by neighboring cells (which may not necessarily be macrophages), without invoking an inflammatory response in the tissue.

Perhaps the most characteristic event of apoptosis is activation of endonuclease(s), which initially cleaves DNA onto 300 and 50 kb sections, and subsequently (although not in all cell types) degrades DNA at the internucleosomal (linker) sites (Arends *et al.*, 1990; Compton, 1992; Wyllie, 1985, 1992; Wyllie *et al.*, 1980). As a result, the products of DNA degradation are the size of nucleosomal and oligonucleosomal DNA fragments which during gel electrophoresis are revealed by a characteristic "ladder" pattern. Apoptotic cells, thus, contain two pools of DNA: the already degraded, low molecular weight DNA of the approximately 180 bp (and its multiples) size, which is relatively easily extractable, and nondegraded, large molecular weight DNA, most

NATO ASI Series, Vol. H 95
Flow and Image Cytometry
Edited by Alain Jacquemin-Sablon

likely still associated with nuclear matrix (Arends *et al.*, 1990; Gong *et al.*, 1994; Oberhammer *et al.*, 1993).

Another characteristic feature of apoptosis is the preservation of the structural integrity and some of the plasma membrane function. However, there is a loss of plasma membrane phospholipid symmetry, resulting in the exposure of phosphaditylserine residues at the surface of cell (Koopman *et al.*, 1994). Cellular organelles, including mitochondria and lysosomes may remain preserved and functional at least during the initial stages of apoptosis apoptosis (Arends *et al.*, 1990; Compton, 1992; Wyllie, 1985, 1992; Wyllie *et al.*,1980).

De novo transcription of certain genes is required for the initiation of apoptosis in some cell types. The products of these genes may represent the effectors of apoptosis, the molecules that are needed for the cell to carry on the program of its suicide. Cells of many types, in particular of hematopoetic lineage, however, undergo apoptosis in the presence of actinomycin D or cycloheximide i.e. when the transcription of the new genes is blocked. Such cells ("apoptosis primed") apparently contain all the necessary molecules for execution of the apoptotic program (Wyllie, 1992).

Observations from different areas stimulated interest in apoptosis in oncology.Thus, it has been noticed that this mode of cell death in tumors is triggered by pharmacological doses of anticancer drugs of various classes, radiation or hyperthermia (Bertrand *et al.*, 1991; Del Bino *et al.*, 1992; Hickman, 1992; Kaufmann, 1989). Furthermore, it has been observed that the cell's ability to undergo apoptosis plays a role in development of B-cell lymphomas (Tsujimoto *et al.*, 1987), as well as, the frequency of spontaneous apoptosis is associated with progression of tumor malignancy (Allan *et al.*, 1992; Hercbergs and Leith, 1993). It has been also seen that the inherent capacity of tumor cells to respond by apoptosis correlates with the expression of several oncogenes or tumor suppressor genes such as *bcl*-2, c-*myc*, *ras*, or p53, and may be prognostic of the treatment (Askew *et al.*, 1991; Ewan *et al.*, 1992; Wyllie *et al.*, 1987). Research of many laboratories is focused on finding the means to modify the responsiveness of normal and tumor cells to various agents by apoptosis. It is expected that a knowledge of the molecular mechanisms of apoptosis will be helpful in developing new antitumor strategies.

While an active participation of the affected cell, often involving *de novo* protein synthesis, is an essential feature of apoptosis, necrosis is a passive, degenerative and catabolic process. Necrosis generally represents a cell's response to gross injury and is frequently induced by an overdose of cytotoxic agents. The early event of necrosis is swelling of cell mitochondria, followed by rupture of the plasma membrane and release of the cytoplasmic contents (Wyllie, 1985). Because the cell constituents, which include proteolytic enzymes, are released from the cell to the tissue during necrosis, this process, in contrast to apoptosis, triggers an inflammatory reaction in the tissue. This often leads to scar formation. DNA degradation is not so extensive during necrosis as in the case of apoptosis, and the products of degradation are heterogenous in size, not forming any discrete bands on electrophoretic gels.

It should be stressed that the classical features of either apoptosis or necrosis are not apparent in in every case of cell death. Numerous examples have been described when the pattern of morphological and/or biochemical changes of the dying cell neither resembled typical apoptosis nor necrosis, but exhibited some features of both (Akagi *et al.*, 1995; Cohen *et al.*, 1992; Collins *et al.*, 1992; Mutsabara *et al.*, 1994; Shinohara and Nakano; Sun *et al.*, 1994; 1993; Zakeri *et al.*, 1993). In some cases, the integrity of the plasma membrane was preserved but DNA degradation was random, without evidence of the internucleosomal cleavage. In other situations, DNA degradation was typical of apoptosis but nuclear fragmentation and other features of apoptosis were not apparent. Generally, while most hematopoietic lineage cell types are "primed" to apoptosis and their death has typical features of apoptosis, the death of epithelial type cells is more complex and sometimes difficult to classify.

The names *delayed reproductive-* or *mitotic- death* are used to describe cell death that occurs as a result of exposure to relatively low doses of drugs or radiation, which induce irreparable damage, but allow cells to complete one or more rounds of cell division (Akagi *et al.*, 1995; Matsubara *et al.*, 1994). Some features of the reproductive cell death resemble apoptosis. It is likely that "reproductive cell death" is actually apoptosis that is delayed and whose picture is complicated by the secondary changes in cell metabolism related to growth imbalance, perturbed cell cycle progression or specific effects of the drug.

The morphological and functional features of cells undergoing apoptosis or necrosis described above provided basis for development of many flow cytometric techniques. Several of these methods are briefly discussed below.

Changes in light scattering properties of the dying cell

Analysis of light scattered by the cell which moves through the laser beam provides information about its size and internal structure. The intensity of light scattered at a forward direction generally correlates with cell size. On the other hand the intensity of scattered light measured at a right angle to the beam (side scatter) correlates with granularity, refractiveness and ability of the intracellular structures to reflect the light (Melamed *et al.*,1990). It could be expected that morphological changes accomanying apoptosis or necrosis, such as cell swelling or shrinkage, breakage of plasma membrane, chromatin condensation, nuclear fragmentation and shedding of apoptotic bodies, alter the cell's ability to scatter light in both, forward and right angle direction.

Indeed, cell necrosis is characterized by a rapid decrease in the cell's ability to scatter light simultaneously in the forward and right angle direction. Most likely, this is a reflection of a plasma membrane rupture and leakage of the cell's constituents. During apoptosis, on the other hand, the decrease in forward light scatter is not initially paralleled by a decrease in side scatter. Actually, an increase in right angle scatter simultaneous with a decrease in forward scatter can be seen in some cell systems (Swat *et al.*, 1991) This may reflect an increased light reflectiveness by condensed chromatin and fragmented nuclei. In later stages of apoptosis, however, the intensity of light scattered at forward as well as at right angle is decreased (Darzynkiewicz *et al.*, 1992).

The assay of cell viability by light scatter analysis is simple and can be combined with the analysis of surface immunofluorescence e.g. to identify the phenotype of the dying cell. It can also be combined with functional assays such as mitochondrial potential, lysosomal proton pump, exclusion of PI or plasma membrane permeability to such dyes as Hoechst 33342 (HO324).

It should be stressed, however, that the light scatter changes cannot be considered specific to apoptosis. Mechanically broken cells, isolated cell nuclei, cell debris and necrotic cells, all have diminished light scatter properties compared to live cells. The light scatter based assays of cell death, therefore, should be accompanied by more specific probes, such that can discriminate between apoptosis and necrosis.

Functional integrity of the plasma membrane.

Loss of the transmembrane transport function and often the loss of the structural integrity of the plasma membrane, are the major features distinguishing dead cells. Numerous assays of cell viability are based on analysis of the plasma membrane function. Because the intact membrane of live cells excludes charged dyes such as trypan blue or propidium iodide (PI), short incubation with these dyes results in selective labeling of dead cells, while live cells show no, or minimal dye uptake (Jacobs and Pipho, 1983; Wallen *et al.*, 1983). The PI exclusion analysis is the most common flow cytometric test of cell viability. In this rapid (5 - 10 min) assay, cell incubation in the presence of 10 - 20 μg/ml of PI labels dead cells, that cannot exclude the dye.

Another assay of membrane integrity employs the nonfluorescent esterase substrate, fluorescein diacetate (FDA). This substrate, after being taken up by live cells, is hydrolyzed. The product of the hydrolysis, fluorescein, is highly fluorescent, emitting green fluorescence (Rotman and Papermaster, 1966). Incubation of cells in the presence of both PI and FDA labels live cells green (fluorescein) and dead cells red (PI). This is a convenient assay, widely used in flow cytometry.

It should be stressed that because during early stages of apoptosis the cell's plasma membrane integrity is still preserved, the early apoptotic cells are undistinguishable from live cells by the tests that utilize exclusion of PI or FDA hydrolysis.

Unlike PI, uncharged fluorochromes that bind to DNA, such as HO342, are not excluded by live or apoptotic cells. However, it has been noticed that whereas short exposure to low concentrations of HO342 led to strong labeling of apoptotic cells, live cells required longer incubation to obtain a comparable intensity of fluorescence (Dive *et al.*, 1992; Ormerod *et al.*, 1992). Supravital uptake of HO342 combined with exclusion of PI (the latter to identify necrotic and late apoptotic cells) and with analysis of the cell's light scatter properties has been proposed by these authors as an assay of apoptosis. It was recently noted that another Hoechst dye, namely Hoechst 33258 has some advantages over HO342 (Elstein and Zucker, 1994). On the other hand substitution of PI by 7-aminoactinomycin D, which fluoresces at higher wavength than PI, makes it possible to combine this assay of apoptosis with a two-color analysis of cell surface immunofluorescence (Schmid *et al.*, 1992).

While the assay based on HO342 uptake has been widely used to identify apoptotic thymocytes, its utility to analyze of apoptosis in other cell systems has yet to be proven. It is unknown whether the increased uptake of HO342 during apoptosis is due to the increased plasma membrane permeability to this dye, or is a result of impairment of the apoptosis-associated P-glycoprotein drug efflux pump, which is known to regulate the extent of HO342 uptake. If the latter is the case, the method may be not be applicable to cells differing in expression or activity of P-glycoprotein.

Expression of phosphatidylserine on surface of plasma membrane during apoptosis.

Phospholipids of the plasma membrane of live cells are asymmetrically distributed across the membrane. Phosphatidylcholine and sphingomyelin are preferentially located on the external leaflet, while phosphatidylserine on the inner leaflet (Op den Kamp, 1979). It has been recently reported by Fadok *et al.*, (1992) that induction of apoptosis of lymphocytes is accompanied by a loss phospholipid asymmetry, resulting in an exposure of phospatidylserine on the outer leaflet. The presence of phosphatidylserine on the cell surface provides a signal for specific recognition of apoptotic cells (apoptotic bodies) by macrophages, thereby facilitating their removeal. Because the anticoagulant annexin V binds with high affinity to the cationic phospholipids such as phosphatidylserine (Andree *et al.*, 1990), one could expect that this ligand may preferentially target apoptotic cells. Indeed, it was recently reported by Koopman *et al.*, (1994) that apoptosis of Burkitt lymphoma cell lines induced by serum deprivation and of germinal B lymphocytes can be flow cytometrically quantified by labeling the cells with fluoresceinated annexin V.

It is unclear, at present, as to whether binding of annexin V is specific to apoptosis, or also characterizes necrotic cells. It is also uknown whether apoptosis in cell systems other than lymphocytes, or induced by antitumor drugs, also can be probed by annexin V. The methodology, however, is very attractive, in particular because cell labeling with annexin V can be combined with cell immunotyping, using fluorochromes of another color than the one conjugated to annexin V.

Functional tests of cell organelles

Some assays of cell viability rely on the functional tests of cell organelles. For example, the charged cationic dye rhodamine 123 (Rh 123) which fluoresces green, accumulates in mitochondria of live cells as a result of the mitochondrial transmembrane potential (Johnson *et al.*, 1980). Cell incubation with Rh 123 results in labeling of live cells while dead cells show minimal Rh 123 uptake. Cell incubation with both Rh 123 and PI labels live cells green (Rh 123) and dead cells red (PI) (Darzynkiewicz *et al.*, 1982). A transient phase of cell death, however, is observed, when the cells partially loose ability to exclude PI and yet stain intensively, even more than intact and live cells, with Rh 123. This may suggest that concomitant with a loss of plasma membrane function the mitochondrial transmembrane potential may be transiently increased (Darzynkiewicz *et al.*, 1982).

Another organelle which can be probed in cell viability tests are lysosomes. Incubation of cells in the presence of 1-2 μg/ml of the metachromatic dye acridine orange (AO) results in the uptake of this dye by lysosomes of live cells which fluoresce red (Traganos *et al.*, 1994). The uptake is the result of an active proton pump in lysosomes: the high proton concentration (low pH) causes AO, which can enter the lysosome in the uncharged form, to become protonated and thus entrapped in the organelle. Dead cells, at that low AO concentration, exhibit weak green and minimal red fluorescence. This assay is especially useful for cells that have numerous active lysosomes, such as monocytes, macrophages, *etc.*.

It should be stressed here again that because in the early stages of apoptosis both the plasma membrane is preserved and most organelles and cellular functions remain relatively unchanged compared to live cells, the assays based on functional assays of the cell organelles, similarly as the dye exclusion tests, are nor able to identify cells that die by apoptosis.

Chromatin condensation during apoptosis

Extensive chromatin condensation is an early event of apoptosis. It is also one of the most characteristic events of apoptosis. The method of detection of chromatin condensation is based on the earlier observation that the sensitivity of DNA *in situ* to denaturation is higher in condensed chromatin of mitotic cells compared to the noncondensed chromatin of interphase cells (Darzynkiewicz *et al.*, 1987). A flow cytometric method for measurement DNA *in situ* sensitivity to denaturation is based on the metachromatic property of the dye AO. This dye can differentially stain double stranded (ds) *vs.* single stranded (ss) nucleic acids (Darzynkiewicz *et al.*, 1987). Specifically, when AO intercalates into ds DNA it emits green fluorescence. In contrast, the product of AO interaction with ss DNA fluoresces red. In this method the cells are briefly pre-fixed in formaldehyde followed by ethanol postfixation. Cell fixation abolishes staining of lysosomes, as described above in this article. Following removal of RNA from the fixed cells by their preincubation with RNase, DNA is partially denatured *in situ* by short cell exposure to acid. The cells are then stained with AO at low pH to prevent DNA renaturation. Apoptotic cells, having a greater proportion of DNA in the denatured form, have higher component of red- and reduced component of green- fluorescence, compared to nonapoptotic (interphase) cells. The latter stain strongly green but have low red fluorescence (Hotz *et al.*, 1992).

Because identification of apoptotic cells by the the above method relies on changes in chromatin structure (chromatin condensation) and not on activation of the endonuclease, the method may be uniquely applicable in situations when internucleosomal DNA degradation is not apparent (Cohen *et al.*, 1992; Collins *et al.*, 1992; Zakeri *et al.*, 1993), i.e. when the methods based on detection of DNA degradation cannot be used. Unfortunately, however, the method does not fully discriminate between mitotic and apoptotic cells. Furthermore, DNA in necrotic cells also appears to be sensitive to denaturation (Darzynkiewicz *et al.*, 1992).

Extraction of degraded DNA from apoptotic cells

As mentioned, extensive DNA cleavage is a very characteristic feature of apoptosis (Arends *et al.*, 1990; Compton, 1992). This feature provides a basis for the development of the flow cytometric assays to distinguish apoptotic cells. Two approaches are frequently used to detect DNA cleavage. One is based on extraction of low MW DNA prior to cell staining, the other on labeling DNA strand breaks.

In the first approach the cellular DNA content is measured following cell permeabilization with detergents or prefixation with precipitating fixatives such as alcohols, or acetone. Cell permeabilization or alcohol fixation does not fully preserve the degraded DNA within apoptotic cells: this fraction of DNA is extracted during subsequent cell rinsing and staining. In consequence, apoptotic cells contain reduced DNA content and therefore can be recognized as cells with low DNA stainability ("sub-G_1"peak), below that of G_1 cells (Nicoletti *et al.*, 1991; Ojeda *et al.*, 1992; Telford *et al.*, 1991; Umansky *et al.*, 1981).

It should be mentioned, however, that the degree of DNA degradation varies depending on the stage of apoptosis, cell type and often the nature of the apoptosis-inducing agent. In consequence, the extractability of DNA during the staining procedure (and thus separation of apoptotic from live cells), also varies. It has been noted that addition of high molarity phosphate-citric acid buffer to the rinsing solution enhances extraction of the degraded DNA (Darzynkiewicz *et al.*, 1992; Gong *et al.*, 1994). This approach can be used to modulate the deegree of DNA extraction from apoptotic cells to the desired level in order to obtain optimal separation of apoptotic from live cells by flow cytometry. Furthermore, DNA extracted by this buffer from the the very same cells which are measured by flow cytometry, can be subjected to analysis by gel electrophoresis (Gong *et al.*, 1994)

Measurement of DNA content enables one to estimate the cell cycle position of the nonapoptotic cells. The methods based on DNA extraction, therefore, can be applied to investigate the cell cycle specificity of apoptosis. Another advantage of this approach is its simplicity and applicability to any DNA fluorochrome (Telford *et al.*, 1991).

There are some drawbacks in identification of apoptotic cells by DNA content.One is its the low specificity in terms of the detection of apoptosis. The "sub G_1" peak can represent, in addition to apoptotic cells, mechanically damaged cells (cell debris), cells with lower DNA content (e.g. in a sample containing cell populations with different DNA indices) or cells with different chromatin structure (e.g. cells undergoing erythroid differentiation) in which the accessibility of DNA to the fluorochrome is diminished Darzynkiewicz *et al.*, 1984). Furthermore, in addition to nuclear fragments, individual chromosomes released from mitotic cells or micronuclei, all may be mistakenly identified as apoptotic cells, especially when one is using logarithmic scale of DNA content coordinate to collect and analyze the data. Another problem stems from lysing of the unfixed cells with detergent in hypotonic solutions, as suggested in some protocols (Nicoletti *et al.*, 1992). The lysis results in isolation of nuclei. Because apoptotic cells have fragmented nuclei, a single apoptotic cell may release several individual nuclear fragments. Hence, the number of "sub G_1" cells in such preparation represents the number of nuclear fragments and provides no information on the number of apoptotic cells.

Labeling of DNA strand breaks in apoptotic cells

Another approach towards the detection of the endonucleolytic DNA cleavage in apoptotic cells is based on *the in situ* detection of a massive number of DNA strand breaks. In this procedure the cells are pre-fixed in formaldehyde. Unlike fixation in precipitating fixatives, formaldehyde by crosslinking DNA fragments to neighboring chromatin constituents, prevents the extraction of degraded, low MW DNA. Thus, unless the apoptotic process is much advanced, the DNA content of apoptotic cells, and with it the number of DNA strand breaks, are not markedly diminished compared to unfixed cells with intact plasma membrane. The presence of the 3' hydroxyl termini in the breaks is detected by attaching to them biotin or digoxygenin conjugated nucleotides, in a reaction catalyzed by exogenous terminal deoxynucleotidyl transferase (TdT; end labeling (Gorczyca *et al.*, 1992; 1993a; 1993b) or DNA polymerase (nick translation) (Gorczyca *et al.*, 1993c; Meyaard *et al.*,1992). Fluorochrome conjugated avidin or digoxygenin antibody is used in the second step of the reaction to label DNA strand breaks. A simplified, single step procedure has been recently developed, using deoxynucleotides directly conjugated to fluorochromes (Li *et al.*, 1995). Commercial kits designed to label DNA strand breaks for identification of apoptotic cells are offered by ONCOR Inc., (Gaithersburg, MD, USA; double step assay) and Phoenix Flow Systems (San Diego, CA, USA, single step assay).

The approach based on DNA strand break labeling appears to be the most specific in terms of positive identification of apoptotic cells. Namely, necrotic cells or cells with primary breaks induced by X-rays irradiation (up to the dose of 25 Gy), or DNA damaging drugs, have an order of magnitude fewer DNA strand breaks than apoptotic cells (Gorczyca *et al.*, 1992). Because cellular DNA content of not only nonapoptotic but also apoptotic cells is measured, the method offers a unique possibility to analyze the cell cycle position, and/or DNA ploidy, of apoptotic cells Gorczyca *et al.*, 1993a; 1993b). This approach can be also combined with analysis of DNA replication (BrdUrd incorporation) utilizing the recently developed SBIP (DNA strand break induction by photolysis) methodology (Li *et al.*, 1994). The DNA strand break labeling method appears to be particularly useful to clinical material and was applied to measure apoptosis of blast cells in peripheral blood or bone marrow induced during chemotherapy of leukemia (Gorczyca *et al.*, 1993a).

Strategies in analysis of apoptosis or necrosis by flow cytometry

With so many methods to measure cell death that are available, which one to chose? The choice of the optimal method for identification of dead cells depends on the cell system, the nature of death inducer, the mode of cell death, the particular information that is being sought (e.g. specificity of apoptosis with respect to the cell cycle phase or DNA ploidy) and the technical restrictions (e.g. the need for sample transportation, type of flow cytometer available, *etc.*).

In principle, the methods based on analysis of the plasma membrane integrity (exclusion of PI or FDA hydrolysis), although simple, and of low cost, fail to identify apoptotic cells, especially at early stages of apoptosis. They can be used to identify necrotic cells, cells damaged mechanically or cells advanced in apoptosis. Their major application is in enumeration of live cells in cultures (e.g. for analysis of cell growth curves), or in discrimination of the mechanically broken cells (e.g. following cell isolation from solid tumors).

The identification of apoptotic cells is more difficult. The most specific appear to be the methods based on the detection of DNA strand breaks. The number of DNA strand breaks in apoptotic cells is of such large magnitude that the intensity of DNA strand break labeling appears to be a specific marker of these cell (Gorczyca *et al.*, 1992, 1993b). The situation, however, is complicated in atypical apoptosis, which is characterized by the lack of internucleosomal DNA degradation (Cohen *et al.*, 1992; Collins *et al.*, 1992; Shinohara and Nakano, 1993; Sun *et al.*, 1994; Zakeri *et al.*, 1993). The number of DNA strand breaks in such atypical apoptotic cells may be inadequate for their identification by this method. On the other hand false positive recognition of apoptosis, by this method, may happen in situations when internucleosomal DNA cleavage accompanies necrosis (Fukuda *et al.*, 1993).

Identification of a mode of cell death is assured with greater confidence if more than a single viability assay is used. Thus, simultaneous assessment of plasma membrane integrity (e.g. exclusion of PI or hydrolysis of FDA), combined with either membrane permeability (HO342 uptake), binding of annexin V, mitochondrial transmembrane potential (Rh 123 uptake), chromatin condensation or DNA cleavage assays, offer a more certain means of identification of the mode of cell death than each of these methods alone.

It should be stressed, however, that depending on the cell system and nature of the inducer of cell death, the metabolic and even morphological features of the dying cell may differ very much from the classic picture of apoptosis or necrosis such as seen in the case of some cells of the hematopoetic lineage. It is difficult in such cases to ascertain the exact mechanism of cell death.

Regardless of the assay that has been used to identify the mode of cell death, the flow cytometric analysis should always be confirmed by the inspection of cells under light or electron microscope. Morphological changes during apoptosis are unique and they should be deciding when ambiguity arises regarding the mechanisms of cell death.

Acknowledgements

Supported by NCI Grant RO1 CA 29 704.

References

Akagi Y, Ito K and Sawada S (1995) Radiation-induced apoptosis and necrosis in Molt-4 cells: a study of dose-effect relationship and their modification. *Int J Radiat Biol* **64**: 47-56,.

Allan DJ, Howell A, Roberts SA, Williams GT, Watson RJ, Coyne JD, Clarke RB, Laidlaw IJ and Potten CS (1992) Reduction in apoptosis relative to mitosis in histologically normal epithelium accompanies fibrocystic change and carcinoma of the premenopausal human breast. *J Pathol* **167**: 25-32.

Andree HAM, Reutelingsperger CPM, Hauptman R, Hemker HC, Hermens WT, and Wilhelm GM (1990) Binding of vascular aniticoagulant α (VACα) to planar phospholipid bilayers. J Biol. Chem., 265: 4923-4927.

Arends MJ, Morris RG and Wyllie AH (1990) Apoptosis: The role of endonuclease. *Am J Pathol* **136**: 593-608.

Askew DA, Ashmun RA, Simmonds BC and Cleveland JL (1991) Constitutive c-myc and IL-3-dependent myeloid cell line suppresses cell cycle arrest and accelerates apoptosis. *Oncogene* **6**: 1915-1922.

Bertrand R, Kerrigan D, Sarang M and Pommier Y (1991) Cell death induced by topoisomerase inhibitors. *Biochem Pharmacol* **42**: 77-85.

Cohen GM, Su X-M, Snowden RT, Dinsdale D and Skilleter DN (1992) Key morphological features of apoptosis may occur in the absence of internucleosomal DNA fragmentation. *Biochem J* **286**: 331-334, 1992.

Collins RJ, Harmon BV, Gobe GC and Kerr JFR (1992) Internucleosomal DNA cleavage should not be the sole criterion for identifying apoptosis. *Int J Radiat Biol* **61**: 451-453.

Compton MM (1992) A biochemical hallmark of apoptosis: Internucleosomal degradation of the genome. *Cancer Metast Rev* **11**: 105-119.

Darzynkiewicz Z, Bruno S, Del Bino G, Gorczyca W, Hotz MA, Lassota P and Traganos F (1992) Features of apoptotic cells measured by flow cytometry. *Cytometry* **13**: 795-808.

Darzynkiewicz Z, Traganos F, Carter SP and Higgins PJ (1987) *In situ* factors affecting stability of the DNA helix in interphase nuclei and metaphase chromosomes. *Exp Cell Res* **172**: 168-179.

Darzynkiewicz Z, Traganos F, Kapuscinski J, Staiano-Coico L and Melamed MR (1984) Accessibility of DNA in situ to various fluorochromes: Relationship to chromatin changes during erythroid differentiation of Friend leukemia cells. *Cytometry* **5**: 355-363.

Darzynkiewicz Z, Traganos F, Staiano-Coico L, Kapuscinski J and Melamed MR (1982) Interactions of rhodamine 123 with living cells studied by flow cytometry. *Cancer Res* **42**: 799-806.

Del Bino G, Bruno S, Yi, P-N and Darzynkiewicz Z (1992) Apoptotic cell death triggered by camptothecin or teniposide. The cell cycle specificity and effects of ionizing radiation. *Cell Prolif* **25**: 537-548.

Dive C, Gregory CD, Phipps DJ, Evans DL, Milner AE and Wyllie AH (1992) Analysis and discrimination of necrosis and apoptosis (programmed cell death) by multiparametr flow cytometry. *Biochem Biophys Acta* **1133**: 275-285.

Elstein KH and Zucker R (1994) Comparison of cellular and nuclear flow cytometric techniques for discrimination apoptotic subpopulations. *Exp Cell Res* **211**: 322-331.

Evan GI, Wyllie AH, Gilbert CS, Littlewood TD, Land H, Brooks M, Waters CM, Penn LZ and Hancock DC (1992) Induction of apoptosis in fibroblasts by c-*myc* protein. *Cell* **69**: 119-128.

Fadok VA, Voelker DR, Campell PA, Cohen JJ, Bratton DL, and Henson PM (1992) Exposure of phosphatidylcholine on the surface of apoptotic lymphocytes triggers specific recognition and removal by macrophages. *J. Immunol.*, **148:** 2207-2211.

Fukuda K, Kojiro M and Chiu J-F (1993)Demonstration of extensive chromatin cleavage in transplanted Morris hepatoma 7777 tissue: Apoptosis or necrosis? *Am J Pathol* **142**: 935-946.

Gong J, Traganos F nad Darzynkiewicz Z (1994) A selective procedure for DNA extraction from apoptotic cells applicable for gel electrophoresis and flow cytometry. *Anal Biochem* **218:** 314-319.

Gorczyca W, Bruno S, Darzynkiewicz RJ, Gong J and Darzynkiewicz Z (1992) DNA strand breaks occurring during apoptosis: Their early in situ detection by the terminal deoxynucleotidyl transferase and nick translation assays and prevention by serine protease inhibitors. *Int J Oncol* **1**: 639-648.

Gorczyca W, Bigman K, Mittelman A, Ahmed T, Gong J, Melamed MR and Darzynkiewicz Z (1993a) Induction of DNA strand breaks associated with apoptosis during treatment of leukemias. *Leukemia* (Baltimore) **7**, 659-670.

Gorczyca W, Gong J, Ardelt B, Traganos F and Darzynkiewicz Z (1993b) The cell cycle related differences in susceptibility of HL-60 cells to apoptosis induced by various antitumor agents. *Cancer Res* **53**: 3186-3192.

Gorczyca W, Gong J and Darzynkiewicz Z (1993c) Detection of DNA strand breaks in individual apoptotic cells by the in situ terminal deoxynucleotidyl transferase and nick translation assays. *Cancer Res* **52**: 1945-1951.

Hercbergs A and Leith JT (1993) Spontaneous remission of metastatic lung cancer following myxedema coma - an apoptosis-related phenomenon? *J Natl Canc Inst* **85**: 1342-1344.

Hickmann JA (1992) Apoptosis induced by anticancer drugs. *Cancer Metast Rev* **11**: 121-139.

Hotz MA, Traganos F and Darzynkiewicz Z (1992) Changes in nuclear chromatin related to apoptosis or necrosis induced by the DNA topoisomerase II inhibitor fostriecin in MOLT-4 and HL-60 cells are revealed by altered DNA sensitivity to denaturation. *Exp Cell Res* **201**: 184-191.

Jacobs DP, Pipho C (1988) Use of propidium iodide staining and flow cytometry to measure antibody - mediated cytotoxicity. Resolution of complement - sensitive and resistant cells. *J Immunol Meth* **62**: 101-110.

Johnson LV, Walsh ML and Chen LB (1980) Localization of mitochondria in living cells with rhodamine 123. *Proc Natl Acad Sci USA* **77**: 990-994.

Kaufmann SH (1989) Induction of endonucleolytic DNA cleavage in human acute myelogenous leukemia cells by etoposide, camptothecin and other cytotoxic anticancer drugs: A cautionary note. *Cancer Res* **49**: 5870-5878.

Koopman G, Reutelingsperger CPM, Kuijten GAM, Keehnen RMJ, Pals ST and van Oears MHJ (1994) Annexin V for flow cytometric detection of phosphatidylserine expression on B cells undergoing apoptosis. *Blood,* **84:** 1415-1420.

Li X, Traganos F, Melamed MR and Darzynkiewicz Z (1995) A single-step procedure for labeling DNA strand breaks with fluorescein- or BODIPY- conjugated deoxynucleotides: Detection of apoptosis and bromodeoxyuridine incorporation. *Cytometry* **20**: (in press)

Li X, Traganos F and Darzynkiewicz Z (1994) Simultaneous analysis of DNA replication and apoptosis during treatment of HL-60 cells with camtothecin and hyperthermia and mitogen stimulation of human lymphocytes. *Cancer Res* **54**: 4289-4293.

Matsubara K, Kubota M, Kuwakado K, Hirota H, Wakazono Y, Okuda A, Bessho R, Lin, YW, Adachi S and Akiyama Y (1994) Variable susseptibility to apoptosis induced by calcium inonophore in hybridomas between HL-60 promyelocytic and CEM T-lymphocytic leukemia cell lines: Relationship to constitutive Mg^2 -dependent endonuclease. *Exp Cell Res* **213**: 412-417.

Melamed MR, Lindmo T and Mendelsohn ML (eds) (1990) Flow Cytometry and Sorting. Wiley-Liss, New York, 1990

Meyaard L, Otto SA, Jonker RR, Mijnster MJ, Keet, RPM and Miedema F (1992) Programmed death of T cells in HIV-1 infection. *Science* **257**, 217-219.

Nicoletti I, Migliorati G, Pagliacci MC, Grignani F and Riccardi C (1991) A rapid and simple method for measuring thymocyte apoptosis by propidium iodide staining and flow cytometry. *J Immunol Meth* **139**: 271-280.

Oberhammer F, Wilson JM, Dive C, Morris ID, Hickman JA, Wakeling AE, Walker PR and Sikorska M (1993) Apoptotic death in epithelial cells: Cleavage of DNA to 300 and/or 50 kb fragments prior to or in the absence of intenucleosomal fragmentation. *The EMBO J* **12**: 3679-3684.

Ojeda F, Guarda MI, Maldonaldo C and Folch H (1992) A flow-cytometric method to study DNA fragmentation in lymphocytes. *J. Immunol Meth* **152**: 171-176.

Op den Kamp JAF (1979) Lipid asymmetry in membranes. *Annu Rev. Biochem.,* **48:** 47-62.

Ormerod MG, Collins MKL, Rodriguez-Tarduchy G and Robertson D (1992) Apoptosis in interleukin-3- dependent hemopoietic cells: quantitation by two flow cytometric methods. *J Immunol Meth* **153**: 57-66.

Rotman B and Papermaster BW (1966) Membrane poroperties of living cells as studied by enzymatic hydrolysis of fluorogenic esters. *Proc Natl Acad Sci USA* **55**:134-139.

Schmid I, Krall WJ, Uittenbogaart CH, Braun J and Giorgi JV (1992) Dead cell discrimination with 7-aminoactinomycin D in combination with dual color immunofluorescence in single laser flow cytometry. *Cytometry* **13:** 204-208 1992.

Shinohara K and Nakano H (1993) Interphase death and reproductive death of X-irradiated

MOLT-4 cells. *Radiat Res* **135**: 197-205.

Sun DY, Jiang, S, Zheng LM, Ojcius DM and Young JDE (1994) Separate metabolic pathways leading to DNA fragmentation and apoptotic chromatin condensation. *J. Exp. Med.*, **179:** 559-565.

Swat W, Ignatowicz L and Kisielow P (1991) Detection of apoptosis in immature $CD4^{+}8^{+}$ thymocytes. *J Immunol Meth* **137**: 79-87.

Telford WG, King LE and Fraker PJ (1991) Evaluation of glucocorticoid-induced DNA fragmentation in mouse thymocytes by flow cytometry. *Cell Prolif* **24**: 447-459.

Traganos F and Darzynkiewicz Z (1994) Lysosomal proton pump activity: supravital cell staining with acridine orange differentiates leukocyte subpopulations. *Meth Cell Biol* **41**: 185-195.

Tsujimoto Y, Ikegaki N and Croce CM (1987) Characterization of the protein product of bcl-2, the gene involved in human follicular lymphoma. *Oncogene* **2**: 3-7..

Umansky SR, Korol' BR and Nelipovich PA (1981) In vivo DNA degradation in the thymocytes of gamma-irradiated or hydrocortisone-treated rats. *Biochim Biophys Acta* **655**: 281-290.

Wallen CA, Higashikubo R and Rot-Roti JL (1983) Comparison of the cell kill measured by Hoechst - propidium iodide flow cytometric assay and the colony forming assay. *Cell Tissue Kinet* **13**: 357-362.

Wyllie AH (1985) The biology of cell death in tumors. *Anticancer Res* **5**: 131-142.

Wyllie AH (1992) Apoptosis and the regulation of cell numbers in normal and neoplastic tissues: An overview. *Cancer Metast Rev* **11**: 95-103.

Wyllie AH, Kerr JFR and Currie AR (1980) Cell death: The significance of apoptosis. In: G.H. Bourne, F.J. Danielli, and K.W. Jeon (eds.), *International Review of Cytology*, Vol. 68. pp. 251-306, New York: Academic Press, 1980.

Wyllie AH, Rose KA, Morris RG, Steel CM, Foster E and Spandidos DA (1987) Rodent fibroblast tumors expressing human myc and ras genes: Growth, metastasis and endogenous oncogene expression. *Br J Cancer* **56**: 251-259.

Zakeri ZF, Quaglino D, Latham T and Lockshin RA (1993) Delayed internucleosomal DNA fragmentation in programmed cell death. *FASEB J* **7**:470-478.

Molecular Cytogenetics: Uses of flow sorted chromosomes, fluorescence *in situ* hybridisation (FISH) and digital microscopy for the analysis of genomes.

Nigel P. Carter
The Sanger Centre
Hinxton Hall
Hinxton
Cambridge CB10 1RQ
UK.

The application of molecular techniques to the study of chromosome structure and rearrangements has revolutionised cytogenetics. Recombinant DNA technology and flow sorted chromosome libraries now provide us with specific probes which, coupled with *in situ* hybridisation techniques, allow the direct visualisation of the position of DNA sequences on metaphase chromosomes and in interphase nuclei. While these techniques allow direct gene mapping and ordering, their application to clinical samples now enables precise diagnosis of chromosomal abnormalities.

Fluorescence *in situ* hybridisation (FISH)

The most direct form of gene mapping is to visualise directly on a banded metaphase chromosome the hybridisation position of a cloned DNA probe using a technique such as FISH (Trask, 1990). The cloned sequence is modified enzymatically with hapten or fluorochrome conjugated bases to form the probe. The DNA probe is denatured to a single stranded form and hybridised to similarly denatured metaphase chromosomes (see Figure 1). Excess and non-specifically hybridised probe is washed off and the hybridisation focus revealed either by the fluorescence of the probe itself or

NATO ASI Series, Vol. H 95
Flow and Image Cytometry
Edited by Alain Jacquemin-Sablon

after incubation with fluorochrome-conjugated ligand. The simultaneous use of fluorescent banding techniques allows the location of the sequence to be determined to a particular chromosome band by simple observation on the epifluorescence microscope (see Figure 2).

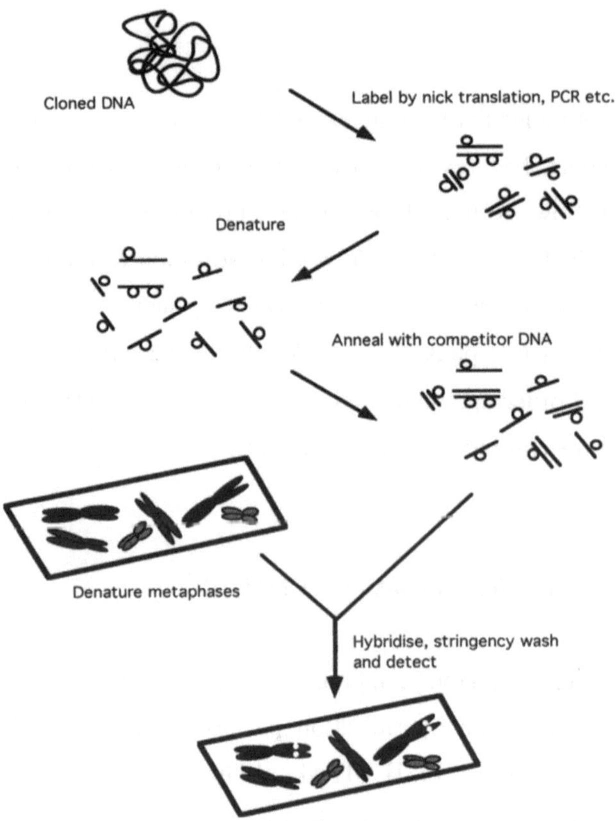

Figure 1. General principles of *in situ* hybridisation

Digital microscopy

The current widespread use of FISH has been due in no small part to the development of improved fluorescence microscopy and digital image capture and analysis systems which allow the reliable capture and enhancement of the often weak fluorescence signals. Commercially produced laser scanning confocal microscopes have been extensively employed for this purpose. However, such microscopes tend to be restricted in the number of fluorochromes that can be used simultaneously due to the limitations of excitation wavelengths produced by lasers.

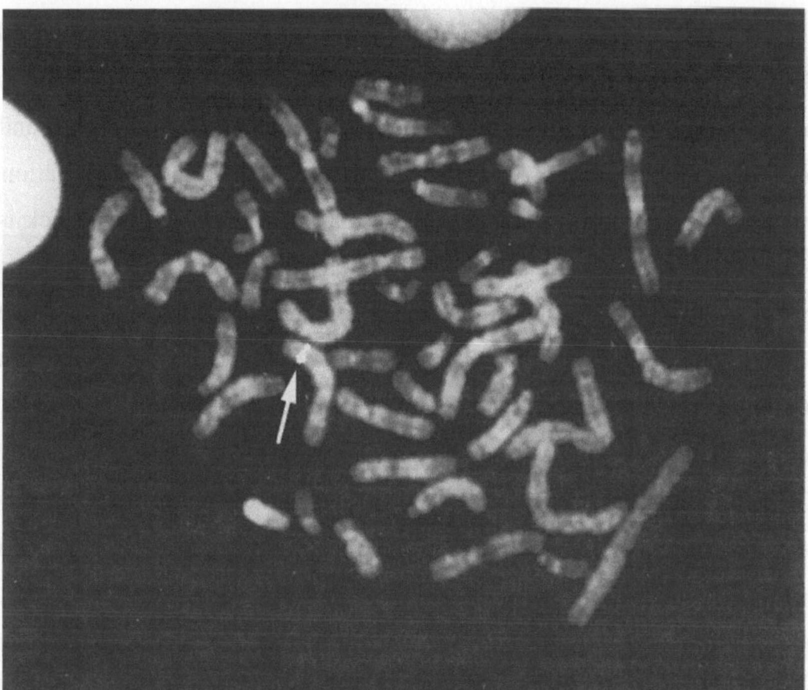

Figure 2. Localisation of a yeast artificial chromosome (YAC) clone onto banded chromosomes to Xq25 (arrowed).

The alternative approach has been to utilise highly sensitive charged coupled device (CCD) solid state cameras with the epifluorescence microscope. The most sensitive scientific grade CCD cameras are monochrome and produce only grey-scale images. However, colour images are constructed by merging three grey-scale images mapped to the red, green and blue planes of a RGB colour image. Thus different filter blocks of the epifluorescence microscope can be used sequentially to produce three grey-scale images which are then merged. Conveniently, it is usual to use blue emitting DNA dyes such as DAPI of Hoechst 33258 to counterstain the DNA and label probes with green emitting FITC and red emitting rhodamine, Texas Red or Cy3. The images acquired from these fluorochromes can be displayed in at least an approximation of true colour on the computer screen.

Unfortunately, changing filter blocks on the microscope inevitably leads to lack of image registration when multiple fluorochromes are superimposed. This problem has been overcome by the development of multiple bandpass filter blocks and excitation filters which allow the simultaneous excitation and viewing of two, three or even four fluorochromes. As the filter block is not moved in order to view the different fluorochromes registration problems are avoided. However, the multiple excitation filter cannot be used with a monochrome camera as the different fluorescent colours would not be distinguished in a single grey-scale image. For monochrome cameras, the multiple excitation filter mounted in the mirror block is removed and individual excitation filters are placed in a computer controlled filter wheel inserted between the lamp housing and the microscope. Under automatic control, the first excitation filter is moved into the excitation light path and a grey-scale image acquired. The second filter is then moved into place and a second image recorded and so on until images for all fluorochromes have been saved into computer memory. The monochrome images are then merged into the false colour image on the computer screen.

Multi-colour FISH

The importance of multi-colour FISH for mapping lies in its ability to order several DNA probes on their parent chromosome (Nederlof *et al.,* 1989, 1990). By utilising different condensation states of chromatin the resolution of gene mapping and ordering can be controlled. In this way the resolution on metaphase preparations is approximately 1 Mb, in interphase nuclei 50 kb (Trask *et al.,* 1989) and on extended chromatin fibres as low as 1 kb (Fidlerova *et al.,* 1994; Wiegant *et al.,* 1992).

FISH is also an important tool for the diagnosis of chromosome aberrations. For prenatal and pre-implantation diagnosis, multicolour FISH enables all of the common viable aneuploidies to be detected simultaneously within a single cell. These techniques are particularly important when the target cells are rare as in pre implantation diagnosis (Griffin *et al.,* 1993) and in the analysis of fetal cells isolated from the maternal circulation (Zheng *et al.,* 1993).

Chromosome sorting

Many of the cloned DNA sequences used as probes for FISH studies are derived from flow-sorted chromosome libraries. The bivariate analysis by flow cytometry of human chromosome suspensions stained with the two fluorochromes, Hoechst 33258 (specificity for AT-rich DNA) and Chromomycin A3 (specificity of GC-rich DNA) enables the majority of the 24 chromosome types to be resolved and sorted (Gray *et al.,* 1979). In the flow cytometer, chromosomes pass sequentially through two spatially separated argon ion laser beams, the first operated using the UV lines (351-364 nm) to excite Hoechst fluorescence and the second operated at 457.9 nm to excite Chromomycin fluorescence. The fluorescence intensity of each chromosome is measured for both dyes independently and correlated on the bivariate flow karyotype where chromosome types are resolved not only by DNA content but also by base pair ratio (see Figure 3).

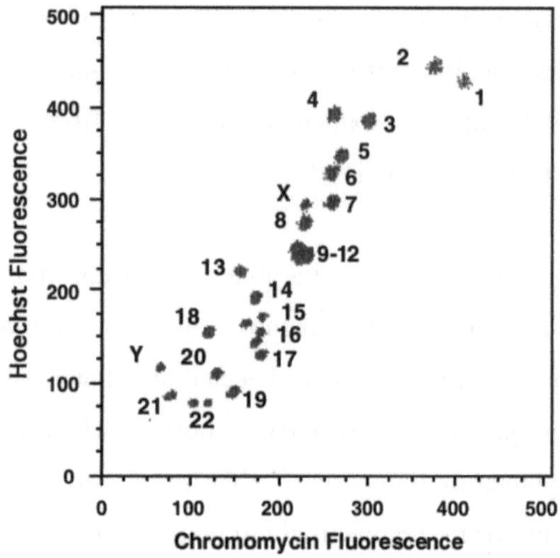

Figure 3. Bivariate flow karyotype of a normal male

With the exception of chromosomes 9-12 which display similar DNA content and base pair ratio, all of the chromosome types can be resolved on the normal bivariate flow karyotype. Utilising the sorting capability of the instrument, fractions highly enriched for a particular chromosome type can be produced for use in the production of DNA libraries, and for cytogenetic analysis of metaphase chromosomes and interphase nuclei. Flow sorted chromosome fractions can also be used for low resolution gene mapping particularly of cloned sequences too small for mapping by FISH. Chromosomes (typically 10,000 chromosomes) can be sorted directly onto filter membranes as small spots of DNA (dot blots) and used directly as targets for isotopic hybridisation (Bernheim *et al.*, 1983). A panel of filters comprising filters for each of the human chromosomes allows the chromosomal location of a particular sequence to be identified.

Many sequences have been localised to specific chromosomes in this way (e.g. Lebo *et al.*, 1984; Collard *et al.*, 1985; Lebo and Bruce 1987; Arai *et al.*, 1989; Neckelmann *et al.*, 1989; Whitcome *et al.*, 1991). However, with the more widespread availability of clone sequence data, gene mapping onto sorted chromosomes can more easily be achieved by direct amplification using the polymerase chain reaction (PCR). Small numbers of chromosomes are sorted directly into PCR tubes to act as the target DNA for amplification (Schmitz *et al.*, 1989; Cotter *et al.*, 1989). As few as 100 chromosomes is sufficient as target for most primer pairs although additional cycles (typically 40) are required for direct detection of product on an agarose gel. A schematic representation of this approach is shown in Figure 4.

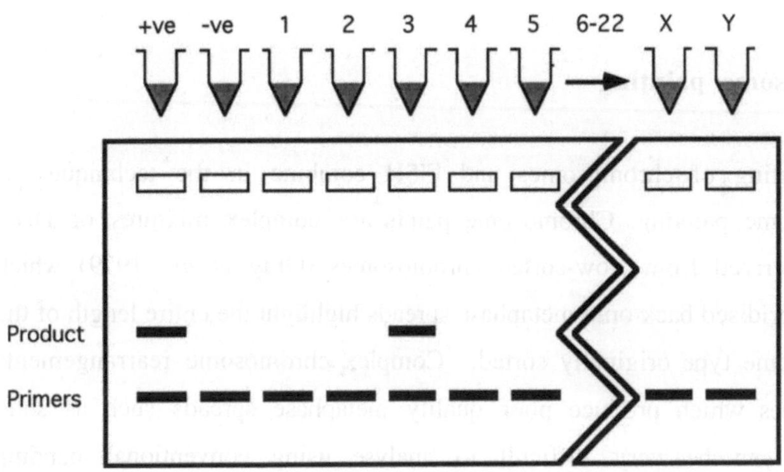

Figure 4. Schematic representation of gene mapping by PCR amplification of sorted chromosomes. In this illustration the sequence is located on chromosome 3.

The resolution of gene mapping by amplification of specific primers from flow sorted chromosomes can be improved by the use of cell lines derived from individuals bearing a balanced translocation. In most cases the two abnormal derivative chromosomes are different in size from the corresponding normal chromosomes and so can be resolved on the bivariate flow karyotype. These derivative chromosomes can thus be sorted into PCR tubes as target DNA for specific amplification. A primer pair which have been previously localised to a particular chromosome can be mapped as being located either distal or proximal to a translocation breakpoint involving that chromosome. By using a panel of cell lines of different translocation breakpoints for the chromosome of interest, it is possible to split the chromosome into a number of intervals only limited by the number of translocation cell lines that can be obtained. We have used this approach to help in the construction of a physical map of the long arm of chromosome 9 (Harris *et al.*, 1993).

Chromosome painting

Flow sorting of chromosomes and FISH combine in the techniques of chromosome painting. Chromosome paints are complex mixtures of DNA probes derived from flow-sorted chromosomes (Gray *et al.*, 1979) which when hybridised back onto metaphase spreads highlight the entire length of the chromosome type originally sorted. Complex chromosome rearrangements and tissues which produce poor quality metaphase spreads such as solid tumours can be very difficult to analyse using conventional banding techniques. Chromosome painting techniques overcome these problems and have now become widely used (Pinkel *et al.*, 1988; Lichter *et al.*, 1988; Cremer *et al.*, 1988; Hulten *et al.*, 1991; Carter 1994). Cloned, flow sorted chromosome libraries have been made in a variety of vectors for all of the human chromosomes and these whole chromosome libraries can be labelled to produce the chromosome paint. The application of polymerase chain reaction technology to the amplification of small numbers of sorted chromosomes enables chromosome paints to be generated rapidly (Telenius *et al.*, 1992a, 1992b) and has allowed the development of reverse chromosome painting where the paint is generated from an aberrant chromosome and then hybridised back onto normal metaphase spreads (Carter *et al.*, 1992). The composition of the aberrant chromosome is thus visualised directly by the pattern of hybridisation on the normal metaphase spreads.

Multicolour FISH techniques are also applicable to chromosome paints. It has even been possible to label up to twelve different chromosome paints with different proportions of multiple fluorochromes allow simultaneous identification of the twelve chromosomes in the same metaphase spread (Dauwerse *et al.*, 1992).

Hybridisation of chromosome paints from one species onto metaphase spreads of another species is also supplying important data of synteny between species and in karyotype evolution (Scherthan *et al.*, 1994). Such comparative genomic hybridisation (CGH) within a species now allows identification of numerical chromosome changes in tissues such as tumours without the need to produce metaphase spreads from the tissue (Kallioniemi *et al.*, 1992). DNA from the tumour is hybridised in one colour with normal DNA in a second colour onto normal metaphase spreads. The fluorescence ratio between the two colours then identifies areas of deletion or amplification in the tumour. This CGH technology is now an important first analysis technique in the study of chromosome rearrangements in tumours and is leading to the identification of new tumour suppressor genes.

References

Arai N, Nomura D, Villaret D, et al. (1989) Complete nucleotide sequence of the chromosomal gene for human IL-4 and its expression. J Immunol 142: 274-82

Bernheim A, Metezeau P, Guellaen G, Fellous M, Goldberg ME, Berger R (1983) Direct hybridization of sorted human chromosomes: localization of the Y chromosome on the flow karyotype. Proc Natl Acad Sci U S A 80: 7571-5

Carter NP (1994) Cytogenetic analysis by chromosome painting. Cytometry 18: 2-10

Carter NP, Ferguson-Smith MA, Perryman MT, et al. (1992) Reverse chromosome painting: a method for the rapid analysis of aberrant chromosomes in clinical cytogenetics. J Med Genet 29: 299-307

Collard JG, de-Boer PA, Janssen JW, Schijven JF, de-Jong B (1985) Gene mapping by chromosome spot hybridization. Cytometry 6: 179-85

Cotter F, Nasipuri S, Lam G, Young BD (1989) Gene mapping by enzymatic amplification from flow-sorted chromosomes. Genomics 5: 470-4

Cremer T, Tesin D, Hopman AH, Manuelidis L (1988) Rapid interphase and metaphase assessment of specific chromosomal changes in neuroectodermal tumor cells by in situ hybridization with chemically modified DNA probes. Exp Cell Res 176: 199-220

Dauwerse JG, Wiegant J, Raap AK, Breuning MH, van-Ommen GJ (1992) Multiple colors by fluorescence in situ hybridization using ratio-labelled DNA probes create a molecular karyotype. Hum Mol Genet 1: 593-8

Fidlerova H, Senger G, Kost M, Sanseau P, Sheer D (1994) Two simple procedures for releasing chromatin from routinely fixed cells for fluorescence in situ hybridization. Cytogenet Cell Genet 65: 203-5

Gray JW, Langlois RG, Carrano AV, Burkhart-Schultz K, Van Dilla MA (1979) High resolution chromosome analysis: one and two parameter flow cytometry. Chromosoma 73: 9-27

Griffin DK, Wilton LJ, Handyside AH, Atkinson GH, Winston RM, Delhanty JD (1993) Diagnosis of sex in preimplantation embryos by fluorescent in situ hybridisation. Bmj 306: 1382

Harris RM, Carter NP, Griffiths B, Goudie D, Hampson RM, Yates JR, Affara NA, Ferguson-Smith MA (1993) Physical mapping within the tuberous sclerosis linkage group in region 9q32-q34. Genomics 15: 265-74

Hulten MA, Gould CP, Goldman AS, Waters JJ (1991) Chromosome in situ suppression hybridisation in clinical cytogenetics. J Med Genet 28: 577-82

Kallioniemi A, Kallioniemi OP, Sudar D, Rutovitz D, Gray JW, Waldman F, Pinkel D (1992) Comparative genomic hybridization for molecular cytogenetic analysis of solid tumors. Science 258: 818-21

Lebo RV, Bruce BD (1987) Gene mapping with sorted chromosomes. Methods Enzymol 151: 292-313

Lebo RV, Gorin F, Fletterick RJ, Kao FT, Cheung MC, Bruce BD, Kan YW (1984) High-resolution chromosome sorting and DNA spot-blot analysis assign McArdle's syndrome to chromosome 11. Science 225: 57-9

Lichter P, Cremer T, Borden J, Manuelidis L, Ward DC (1988) Delineation of individual human chromosomes in metaphase and interphase cells by in situ suppression hybridization using recombinant DNA libraries. Hum Genet 80: 224-34

Neckelmann N, Warner CK, Chung A, et al. (1989) The human ATP synthase beta subunit gene: sequence analysis, chromosome assignment, and differential expression. Genomics 5: 829-43

Nederlof PM, Robinson D, Abuknesha R, Wiegant J, Hopman AH, Tanke HJ, Raap AK (1989) Three-color fluorescence in situ hybridization for the simultaneous detection of multiple nucleic acid sequences. Cytometry 10: 20-7

Pinkel D, Landegent J, Collins C, Fuscoe J, Segraves R, Lucas J, Gray J (1988) Fluorescence in situ hybridization with human chromosome-specific libraries: detection of trisomy 21 and translocations of chromosome 4. Proc Natl Acad Sci U S A 85: 9138-42

Scherthan H, Cremer T, Arnason U, Weier HU, Lima-de-Faria A, Fronicke L (1994) Comparative chromosome painting discloses homologous segments in distantly related mammals. Nat Genet 6: 342-7

Schmitz A, Olschwang S, Chaput B, Thomas G, Frelat G (1989) Oncogene detection by enzymatic amplification on flow sorted chromosomes. Nucleic Acids Res 17: 816

Telenius H, Carter NP, Bebb CE, Nordenskjold M, Ponder BA, Tunnacliffe A (1992a) Degenerate oligonucleotide-primed PCR: general amplification of target DNA by a single degenerate primer. Genomics 13: 718-25

Telenius H, Pelmear AH, Tunnacliffe A, Carter NP, Behmel A, Ferguson-Smith MA, Nordenskjold M, Pfragner R, Ponder BA (1992b) Cytogenetic analysis by chromosome painting using DOP-PCR amplified flow-sorted chromosomes. Genes Chromosom Cancer 4: 257-63

Trask B, Pinkel D (1990) Fluorescence in situ hybridization with DNA probes. Methods Cell Biol 33: 383-400

Trask B, Pinkel D, van-den-Engh G (1989) The proximity of DNA sequences in interphase cell nuclei is correlated to genomic distance and permits ordering of cosmids spanning 250 kilobase pairs. Genomics 5: 710-7

Whitcombe DM, Carter NP, Albertson DG, Smith SJ, Rhodes DA, Cox TM (1991) Assignment of the human ferrochelatase gene (FECH) and a locus for protoporphyria to chromosome 18q22. Genomics 11: 1152-4

Wiegant J, Kalle W, Mullenders L, Brookes S, Hoovers JM, Dauwerse JG, van-Ommen GJ, Raap AK (1992) High-resolution in situ hybridization using DNA halo preparations. Hum Mol Genet 1: 587-91

Zheng YL, Carter NP, Price CM, Colman SM, Milton PJ, Hackett GA, Greaves MF, Ferguson-Smith MA (1993) Prenatal diagnosis from maternal blood: simultaneous immunophenotyping and FISH of fetal nucleated erythrocytes isolated by negative magnetic cell sorting. J Med Genet 30: 1051-6

DNA Topoisomerases as Drug Targets and Cell Cycle Checkpoint Effector Molecules

Paul J. Smith[1] Nicola Blunt and Sylvie Souès
Medical Research Council Clinical Oncology and Radiotherapeutics Unit
MRC Centre
Hills Road
Cambridge
United Kingdom CB2 2QH

Introduction

Both as drug targets and as active participants in cell cycle control, the type II DNA topoisomerases are potentially critical enzymes for determining the chemosensitivity of tumour and normal cells. These nuclear enzymes resolve topological and conformational changes in DNA (Liu, 1989; Wang, 1985) through highly co-ordinated processes of strand cleavage and strand passing (Fig 1). The DNA topoisomerases occupy an important position in anticancer strategies principally because several classes of active anticancer agents can either trap topoisomerase molecules in complexes with DNA or otherwise disturb enzyme function (Fig 1). Enzyme trapping or dysfunction can lead to the disruption of cell cycle progression as a consequence of 'checkpoint' activation and the generation of intracellular signals that lead to apoptosis or prolonged cytostasis (Smith, 1990). An understanding of how anticancer agents target topoisomerases, together with an appreciation of the factors controlling cycle arrest, is vital to the rational use of topoisomerase inhibitors in therapeutic modalities. More importantly elucidating the interplay of the topoisomerase isozymes in drug responses is vital if such agents are to be combined with novel therapeutic strategies aimed at the re-establishment of checkpoint or apoptotic controls in tumours. This brief overview highlights the forms and roles of the mammalian type II topoisomerases together with the importance of cell cycle checkpoints in anticancer drug action. Specific examples are provided for the application of cytometric methods in analysing cellular responses to the novel topoisomerase inhibitors idarubicin and the bis(2,6-dioxopiperazine) derivative ICRF-193.

---

[1] Current Address: Department of Pathology, University of Wales College of Medicine, Heath Park, Cardiff, UK.

NATO ASI Series, Vol. H 95
Flow and Image Cytometry
Edited by Alain Jacquemin-Sablon

Topoisomerases - Forms and Locations:

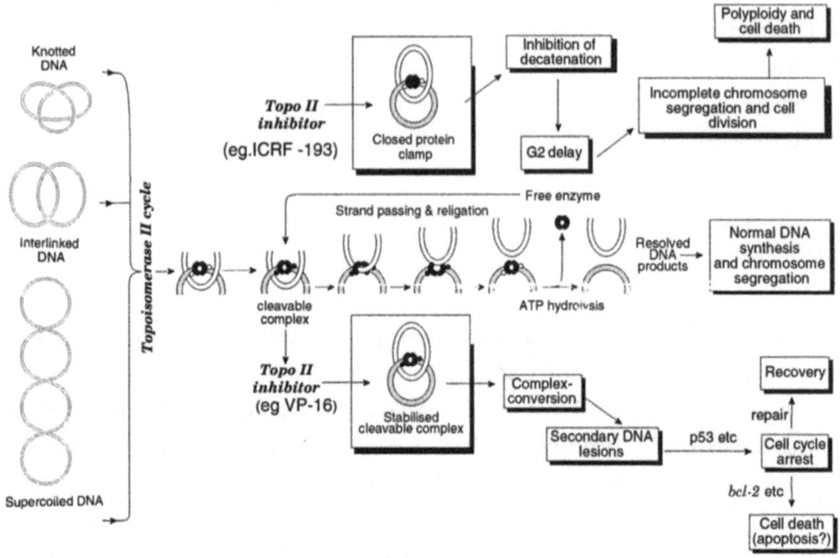

Fig 1 Diagrammatic representation of the normal type II topoisomerase cycle, resolving topological problems in DNA structures. The mechanisms of topoisomerase II inhibition and trapping by anticancer agents are show together with the subsequent cellular responses.

Unlike the type I (single strand-cleaving) enzyme, the type II (double strand-cleaving) enzyme, topoisomerase II, is essential for cell viability facilitating the resolution and separation of daughter chromosomes at mitosis. Differential expression of the type I and II enzymes in tumour and normal tissues implies that selective toxicity could be achievable with type-specific inhibitors. The human DNA topoisomerase II enzyme is found in two forms coded for by single copy genes yielding homodimeric proteins of 170 kDa (topoisomerase IIα) and 180 kDa (topoisomerase IIβ) (Tan et al, 1992). The more abundant isozyme, topoisomerase IIα, is located at nuclear matrix-associated regions inferring both structural and functional roles (Ura and Hirose, 1991; Wood and Earnshaw, 1990). Topoisomerase IIß is found predominantly in the nucleolus (Negri et al 1992) and binds to GC-rich sequences present in gene promotor regions suggesting participation in gene expression (Tan et al, 1992). Fig 1 shows the consequences of either trapping topoisomerases with specific 'poisons' (eg VP-16 and doxorubicin) or preventing decatenation by a potent inhibitor (eg ICRF-193)

Cell Cycle Regulation, Checkpoints and DNA Damage

The major cell cycle transitions are controlled by the cyclin-dependent kinases (CDKs) through systems which co-ordinate the sequence of events and their delay under suboptimal growth conditions. The major mechanisms for CDK regulation can involve cyclin binding, cyclin-mediated substrate-targeting, changes in phosphorylation status and the action of inhibitory proteins (CKIs) that bind and inactivate CDK-cyclin complexes. Mitotic (eg B1), S/G2 (eg A) and G1 (eg C, D1, D2, D3 and E) cyclins have been described, based upon patterns of expression and protein kinase activation. Using flow cytometry to map cell cycle distribution and status of cyclins is one approach for evaluating cellular capacity and 'intent' to traverse the cell cycle.

The ability to arrest at defined checkpoints in the cell cycle when exposed to genotoxic agents, is now recognised (Lane, 1992; Kastan et al., 1992) as an important cellular strategy - reducing the opportunity for damaged DNA to be replicated or for damaged chromosomes to be committed to segregation - allowing time for DNA repair and the expression of recovery pathways.

The effective linking of cycle progression, apoptosis and DNA repair acts to protect against the transmission of genomic instability (Smith et al, 1994). Arrest at G1/S or G2/M are typical but variable responses of proliferating human cells to DNA damaging agents. Dysfunction of p53-dependent pathways can allow tumour cells to evade both arrest at the G1/S checkpoint and apoptosis, thereby gaining a selective advantage in attempting to overcome chemotherapy (Clarke et al, 1993; Livingstone et al, 1992; Lowe et al, 1993). Recent studies have also revealed participation of p53 in G2/M arrest (Stewart et al, 1995), the operation of a mitotic spindle checkpoint (Cross et al, 1995) and in the activation the Gadd45 gene product involved in DNA repair (Smith et al, 1994). p21 (CIP1/WAF1/CAP20/SDI1) is a major CKI which is regulated by transcription and capable of mediating cycle-arrest following its induction through p53 stabilization in response to DNA damage and in senescence. Interestingly, p21 also appears to be capable of differentially affecting the progress of DNA replication and repair through effects on PCNA complexes. The PCNA inhibitory function of p21 is located within the C-terminal domain.

Members of the p21/p27 family of CKIs contain an N-terminal CDK inhibiting domain that can act independently to block cycle progression (Luo et al, 1995).

Checkpoint Evasion and Therapy

Somatic mutation at the p53 locus is a frequent occurrence in human tumours (Hollstein et al, 1991). A key tumour-limiting function of wild-type p53 may be to mediate growth arrest after a given number of cell divisions. Cancers with p53 mutations tend to have a worse prognosis for response to chemotherapy than those showing wild-type alleles (Callahan, 1992). Recent in vivo studies have shown that p53wt cells are more chemoresponsive than p53mut cells and that tumour subpopulations surviving therapy carry p53 mutations (Lowe et al, 1994). Evasion of p53-mediated growth control can involve direct mutation, indirect activation, or 'avoidance' of activation, most probably due to failure to reach a critical threshold of DNA damage. Functionally, that threshold will depend upon the properties of the genotoxic agent, the processing of DNA damage and cellular capacity to signal the presence/persistence of DNA lesions. The induction of p21 and its interactions with other proteins are indicators of the supra-threshold generation of arrest signals and the nature of the effector pathways for arrest. The development of cytometric assays for p21 induction and its interaction with other molecules holds promise for evaluating the responses of tumour cells to anticancer agents. The rate of generation/resolution of DNA lesions, their genomic location, their interactions with replicative and transcriptional machinery, and their visibility to surveillance systems will be critical factors in determining chemosensitivity.

Innate or acquired resistance to topoisomerase-interactive antitumour agents is a significant problem in the chemotherapeutic management of human tumours. There appear to be multiple mechanisms involved, including: the modulation of intracellular drug accumulation, changes in subcellular distribution, enhanced metabolism of drug molecules, reduction of target availability, and gene mutations giving rise to decreased target sensitivity. Checkpoint evasion could also be crucial in providing selective advantage for the generation of multiple resistance mechanisms within a progressive tumour.

DNA Topoisomerases and The Cell Cycle

The level of topoisomerase IIα is cell cycle-regulated, peaking in G2/M and declining to a minimal level at the end of M phase. Flow cytometry has been used to map such regulation and to identify the cell cycle specific modulation of target enzyme availability, and drug sensitivity (Smith, 1990; Smith and Makinson, 1989), by growth factors (Epstein and Smith, 1988). In contrast, the levels of the type I topoisomerase (topoisomerases I) and topoisomerase IIβ are relatively constant throughout the cell cycle. The peaking of topoisomerase IIα at G2/M and the decline thereafter have parallels with the ubiquitin-dependent degradation of cyclins A and B during M phase. Consistent with this is the presence of a cyclin type B destruction box at the N-terminus of human topoisomerase IIα, an incomplete box in topoisomerase IIβ and no box in topoisomerase I. As yet there is no clear consensus on the relevance of the topoisomerase IIβ as a target enzyme for cytotoxicity, whether given tumour types display distinctive patterns of expression (Jenkins et al, 1992) and if topoisomerase IIα destruction is co-ordinated in some functional way with checkpoint traverse and cyclin changes.

Intracellular Localisation of Topoisomerase II-targeting Agents

Drug uptake, export and subcellular distribution are important factors in determining the cytotoxic action of anticancer agents particularly those directed at nuclear-located targets such as topoisomerase II. Even cells of a given tumour type can vary in the extent of drug accumulation as shown by the differences between two human breast tumour cell lines with respect to doxorubicin (DOX) uptake (Fig 2). In this case such differences are not due to the operation of multi-drug transporter resistance mechanisms or attributable to differences in cellular nucleic acid content. Flow cytometry is a valuable tool for quantifying these innate differences and identifying population heterogeneity.

The situation is more complicated when comparing accumulation of different agents in tumour cell lines or biopsies. Idarubicin (IDA) is a novel, topoisomerase II-reactive, anthracycline DNA-intercalator. Drug uptake studies, using flow cytometry (Berman and McBride, 1992) to quantify intracellular anthracycline concentrations, have indicated that the enhanced toxicity of IDA towards multidrug-resistant leukaemia cells was associated with a rapid uptake of IDA compared with daunorubicin. This may arise from differences in the subcellular distribution of drug molecules and their availability for critical target interaction or cellular exclusion mediated via multidrug-resistance transporter pathways.

The natural fluorescence of anthracycline drugs has enabled uptake studies, using high-resolution laser scanning confocal fluorescence microscopy to be performed on intact, viable cells (Smith et al, 1992). Confocal fluorescence images showed intense cytoplasmic labelling of intact, adherent breast tumour cells with IDA compared to DOX. The use of confocal imaging allows a semi-quantitative comparison to be made of cytoplasmic and nuclear uptake within a defined equatorial section of an intact cell (Fig 3). Image analysis shows that the two cell lines studied have similar mean areas for whole cell (650-750 μm^2) and nuclear (230-290 μm^2) sections. Intensity measurements were carried out on randomly selected cells and the DOX- and IDA-associated fluorescence values adjusted for the areas of the regions sectioned. Thus fluorescence intensity per μm^2 provides an indication of drug concentration within a cellular compartment.

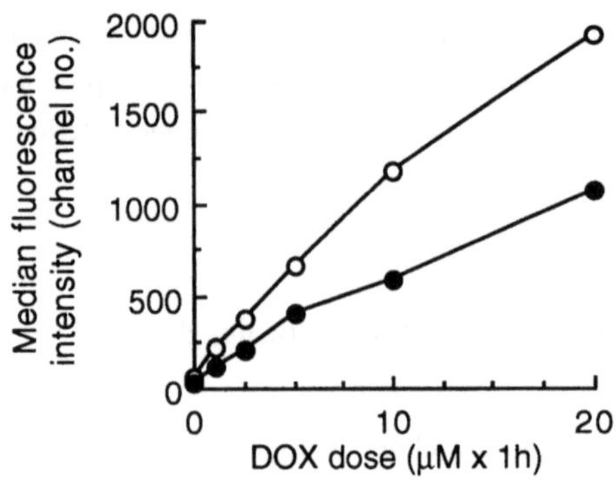

Fig 2 Flow cytometric determination of cellular accumulation of DOX by breast tumour cell lines. Symbols: o, MCF-7; ●, T47D.

Both cell lines showed similar patterns of distribution of DOX with 60-80 % of total fluorescence being associated with the nucleus and the nuclear membrane. The fluorescence intensity per μm^2 of optical section for IDA is 30-60 fold higher than for DOX, the distribution being biased towards the cytoplasm. The T47D cell line appears to accumulate higher concentrations of IDA within both the cytoplasm and the nucleus when compared with the MCF-7 cell line. The results imply that the majority of the fluorescence signal obtained from IDA-treated cells is cytoplasmic whereas it is nuclear-located for DOX-treated cells. Even when differences in fluorescence potential and quenching upon binding to DNA are taken into account, it is clear that IDA is preferentially accumulated within the cytoplasmic compartment of cells compared with DOX. Importantly nuclear-associated fluorescence in IDA-treated cells remains 12-17-fold greater than in DOX-treated cells suggesting at least a 4-fold greater level of delivery of IDA to the nucleus. The relative difference between the two agents was maintained at lower drug concentrations (data not shown) suggesting that the results obtained with the high-dose acute-exposure conditions reveal true differences in the uptake patterns of the two drugs.

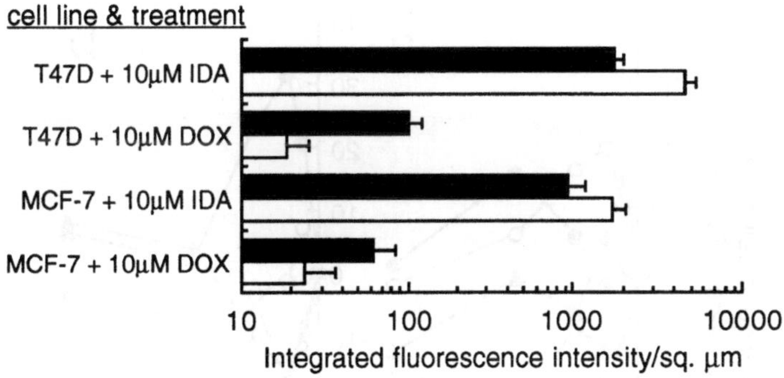

Fig 3 Determination of the distribution of intracellular fluorescence by image analysis of confocal sections of drug-treated (1h exposure) breast tumour cells. Open columns, cytoplasm; closed columns, nucleus.

Cell Cycle Arrest Induced By DOX And IDA In Breast Tumour Cell Lines

Using arrest by acute X-irradiation at either G1/S or G2 as an indicator of cell cycle checkpoint status (Kastan et al., 1992), we have confirmed the arrest status of two breast tumour cell lines: one lacking stringent G1/S checkpoint control (T47D; p53mut) in response to DNA damage and a second (MCF-7; p53wt) which shows active arrest at G1/S following damage induction. Here we have used these checkpoint incompetent and competent human tumour cells to establish a relationship between drug uptake, checkpoint activation and cell cycle arrest.

Fig 4 shows the comparative ability of DOX or IDA to induce cell cycle arrest at the G1/S boundary or G2 in MCF-7 and T47D cell lines exposed to the agents for 24 h. MCF-7 shows a loss of cells from S phase, a dose-dependent G2 block (maximal at 0.05 and 0.2 µM DOX and IDA respectively) and an accumulation of cells in G1 until G2 arrest reduces G1 resupply at higher drug doses. IDA is approximately 4-fold more potent than DOX for the induction of G2 arrest and G1 accumulation in the MCF-7 cell line in keeping with the preferential nuclear targeting. At low DOX concentrations (0.025-0.05 µM) T47D cells show an increase in G2, replaced at higher doses with a dose-dependent delay in S phase. Even at the lowest IDA concentration studied S/G2 progression is severely limited in T47D cells. The effects of either agent on G1 emptying reveals the operation of a mechanism for G1/S arrest in MCF-7 but not in T47D, consistent with the p53 genotypes. The differential G1/S arrest effects of IDA in these asynchronous populations is less obvious than for DOX due to the effects of IDA on S phase progression in T47D.

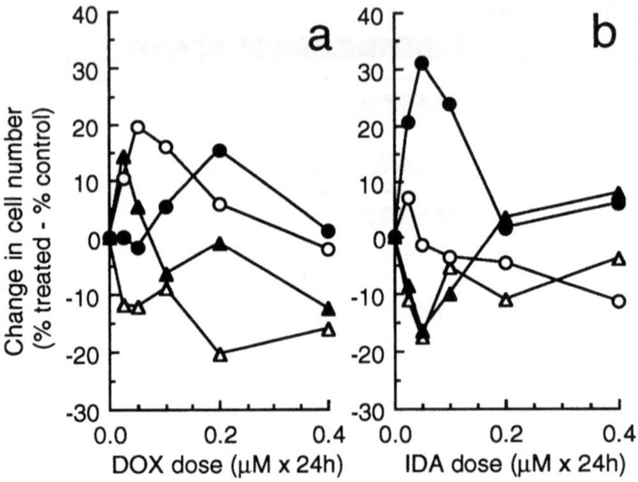

Fig 4 a and b Flow cytometric analysis of drug-induced cell cycle perturbations in MCF-7 (circles) and T47D (triangles) breast tumour cell lines. Symbols: ○ , △ , G1; ● , ▲ , G2/M.

A Topoisomerase II-Dependent Cell Cycle Checkpoint

Topoisomerases appear to have a direct role in cycle regulation, 'sensing' topological (catenation) problems in chromatin and controlling a late G2 'checkpoint' dependent upon topoisomerase II activity (Downes et al, 1994). This checkpoint role has been revealed using a new class of anticancer topoisomerase inhibitors, the bisdioxopiperazines, as enzyme probes. The bisdioxopiperazine derivatives (eg ICRF-193) are inhibitors of mammalian type II DNA topoisomerase and act to uncouple chromosome dynamics from other cell cycle events. Mechanistically ICRF-193 acts by trapping the enzyme in the form of a closed protein-clamp (Ishimi et al, 1992) (Fig 1). The result of this uncoupling of chromosome segregation and cell division results in the progression of cells into a polyploid cycle as shown in the flow cytometric bivariate plots of DNA content versus 90° light scatter. ICRF-193 is a probe for the involvement of topoisomerase II activity in cell cycle transitions. Potentially the agent could be used to uncover the linkage of cellular stress response systems, such as apoptosis, with the dysfunction of a specific G2 checkpoint without the apparent requirement for DNA damage induction.

The extent to which delay is induced at the G2-catenation checkpoint by ICRF-193 can be determined by several techniques. One approach is to follow the progression of synchronised cells through G2/M by flow cytometry. Fig 5 shows that following release of synchronised cells from G1/S blockade, the G2/M fraction of control cultures increases over a 12 h period as the synchronised cohort passes through M. ICRF-193 does not affect the initial rate of accumulation compared with the control but increases the number of cells recruited into G2/M prior to passage of the cohort through M inferring a delay in G2. In control cultures G1 empties out rapidly with evidence of resupply of occurring within 6 h of release from synchrony. Lack of resupply of G1 in ICRF-193-treated cells is due to the passage of G2-delayed cells into a polyploid cycle.

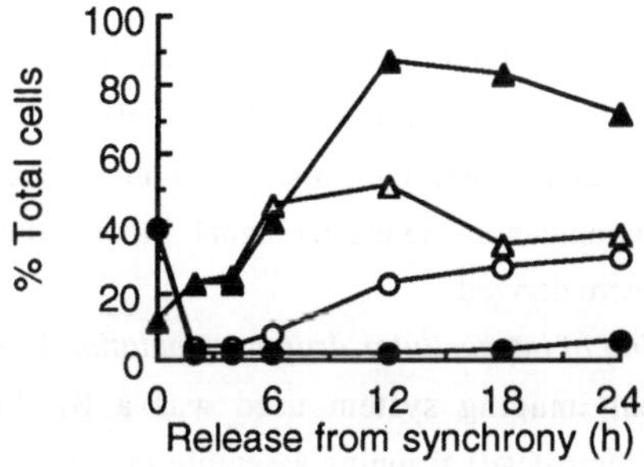

Fig 6 Flow cytometric analysis of cell cycle perturbations induced by ICRF-193 in cultures of SV40 transformed fibroblasts released from synchrony at G1/S (induced by 24h exposure to 1 µg/ml aphidicolin) into normal medium (open symbols) or medium supplemented with 2 µg/ml ICRF-193 (closed symbols). Symbols: ○ , ● , G1; △ , ▲ , G2/M (ie 4n content DNA). Data are typical results from a single experiment.

<u>Materials and Methods</u>

Cell culture and drug treatments: The breast tumour cell lines, T47D and MCF-7, were cultured in RPMI 1640 medium and in Dulbecco's MEM (supplemented with 5% foetal bovine serum, glutamine and antibiotics) respectively. The SV40 transformed human fibroblast cell line MRC5CV1 (normal donor)was grown in Eagle's minimum essential medium supplemented with 10% foetal calf serum, 1mM glutamine and antibiotics and incubated at 37°C in an atmosphere of 5% CO_2 in air. Fresh stock solutions of doxorubicin (Adriamycin; DOX) and idarubicin (IDA) were supplied by Farmitalia Carlo Erba and prepared immediately prior to use. ICRF-193 (kindly supplied by Dr AM Creighton, St Bartholomew's Hospital, London) was prepared as a 2 mg/ml stock solution in DMSO.

Cell Cycle Analysis: Triton X-100-permeabilised cells were treated with RNAse, stained with ethidium bromide and analysed for DNA content by flow cytometry as described previously (Smith and Makinson, 1990).

Nuclear morphology: Cytospin slide preparations of drug-treated and control cells (2 x 10^4 cells/0.25 ml culture medium/sample) were made using a cytocentrifuge operating at 600 rpm for 1 min. Air dried samples were fixed with 4% paraformaldehyde in PBS (pH 7.4) at room temperature for 15 min, rinsed in PBS and stained with acridine orange (0.5 µg/ml; ribonuclease digestion 0.2 mg/ml for 10 min at room temperature), rinsed and then mounted with Vectorshield (Vector Laboratories). Specimens were examined and images collected using an MRC600 Confocal Imaging system, with a standard GHS (514 nm excitation) filter block, linked to a Nikon Optiphot II microscope.

Analysis of drug uptake by flow cytometry: After drug treatments H69 cells were resuspended in phosphate buffered saline (PBS) and analysed immediately using a flow cytometer incorporating an Innova 70 argon laser (Coherent) tuned to 488 nm. Signals for forward and 90° light scatter and fluorescence above 630 nm were collected for $1x10^4$ cells using the 90° light scatter parameter as the master signal. Median fluorescence intensities for populations of single cells were derived.

Fluorescence imaging of intracellular drug accumulation by confocal laser scanning microscopy: The confocal imaging system used was a BioRad MRC-600 (BioRad Microscience Ltd, Hemel Hempstead) scanning assembly incorporating a 25 mW argon ion laser and coupled to a Nikon Optiphot II fluorescence microscope with excitation at 488 nm wavelength. Following drug treatments coverslip cultures of breast tumour cells were mounted in PBS on petroleum jelly supports on a microscope slide. Slides were immediately examined under the confocal laser scanning microscope using a GHS filter block, a x60 oil immersion objective lens. Images of the cells were stored after Kalman filtration on optical cartridge and photographs taken subsequently using Ilford FP4 film.

References

Berman E and McBride M (1992) Comparative cellular pharmacology of daunorubicin and idarubicin in human multidrug-resistant leukaemia cells. Blood 79: 3267-3273.

Callahan R (1992) p53 mutations, another breast cancer prognostic factor J Natl Cancer Inst 84: 826-827

Clarke AR, Purdie CA, Harrison DJ, Morris RG, Bird CC, Hooper ML, Wyllie AH (1993) Thymocyte apoptosis inducecd by p53-dependent and independent pathways. Nature 362: 849-852

Cross SM, Sanchez CA, Morgan CA, Schimke MK, Ramel S, Idzerda, RL, Raskind WH, Reid BJ (1995) A p53-dependent mouse spindle checkpoint. Science 267: 1353-1356

Downes CS, Clarke DJ, Mullinger AM, Gimenez-Abian JF, Creighton AM, Johnson RT (1994) A topoisomerase II-dependent G2 cycle checkpoint in mammalian cells. Nature 372: 467-470

Epstein RJ, Smith PJ (1988) Estrogen-induced potentiation of DNA damage and cytotoxicity in human breast cancer cells treated with topoisomerase II interactive antitumour drugs. Cancer Res 48:297-303

Hollstein M, Sidrasnky D, Vogelstein B, Harris CC p53 mutations in human cancers. Science 253: 49-53 1991

Ishimi Y, Ishida R, Andoh T (1992) Effect of ICRF-193, a novel DNA topoisomerase II inhibitor, on simian virus 40 DNA and chromosome replication in vitro. Mol Cell Biol 12: 4007-14

Jenkins JR, Ayton P, Jones T, Davies SL Simmons, DL, Harris AL, Sheer D & Hickson D (1992) Isolation of cDNA clones encoding the ß isoenzyme of DNA topoisomerase II and localisation of the gene to chromosome 3p24. Nucleic Ac Res 20: 5587-5592

Kastan, MB, Zhan, Q, El-Deiry, WS, Carrier, F, Jacks, T, Walsh, WV, Plunkett, BS, Vogelstein, B, Fornace, AJ Jr (1992) A mammalian cell cycle checkpoint pathway utilizing p53 and GADD45 is defective in ataxia-telangiectasia. Cell 71: p 587-597.

Lane, DP (1992) p53, guardian of the genome. Nature 358: p15-16

Liu LF (1989) DNA topoisomerase poisons as anti-tumour drugs. Ann Rev Biochem 58:351-375

Livingstone LR, White A, Sprouse J, Livanos E, Jacks T, Tlsty T (1992) Altered cell cycle arrest and gene amplification potential accompany loss of wild-type p53. Cell 70: 923-935

Lowe SW Bodis S, McClatchey A, Remington L, Ruley HE, Fisher DE, Housman DE, Jacks T (1994) p53 status and the efficacy of cancer therapy in vivo. Science 266: 807-810

Lowe SW, Ruley HE, Jacks T, Houseman DE (1993) p53-dependent apoptosis modulates the cytotoxicity of anticancer agents. Cell 74: 957-968

Luo Y, Hurwitz J, Massague J (1995) Cell-cycle inhibition by independent CDK and PCNA binding domains in $p21^{Cip1}$. Nature 375: 159-161

Negri C, Chiesa R, Cerino A, Bestagno M, Sala C, Zini N, Maraldi NM, Astaldi Ricotti GCB (1992) Monoclonal antibodies to human DNA topoisomerase I and the two isoforms of DNA topoisomerase II: 170- and 180-kDa isozymes ExpCell Res 200: 452-459

Smith ML, Chen I-T, Zhan Q, Bae I, Chen C-Y, Gilmer TM, Kastan MB, O'Connor PM, Fornace AJ (1994) Interaction of the p53-regulated protein Gadd45 with proliferating cell nuclear antigen. Science 266: 1376-1380

Smith PJ (1990) DNA topoisomerase dysfunction: a new goal for antitumor chemotherapy. BioEssays 12:167-172

Smith PJ, Makinson TA (1989) Cellular consequences of overproduction of DNA topoisomerase II in an ataxia-telangiectasia cell line. Cancer Res 49:1118-1124

Smith PJ, Sykes HR, Fox ME, Furlong IJ. (1992) Subcellular distribution of the anticancer drug mitoxantrone in human and drug-resistant murine cells analyzed by flow cytometry and confocal microscopy and its relationship to the induction of DNA damage.Cancer Research. 52: 4000-8.

Stewart N, Hicks GG, Paraskevas F, Mowat M (1995) Evidence for a second cell cycle block at G2/M by p53. Oncogene 10: 109-115

Tan KB, Dorman TE, Falls KM, Chung TD, Mirabelli CK, Crooke ST, Mao J (1992) Topoisomerase IIα and topoisomerase IIß genes: characterization and mapping to human chromosomes 17 and 3, respectively. Cancer Res 52: 231-234

Ura K, Hirose S (1991) Possible role of DNA topoisomerase II on transcription of the homeobox gene Hox-21 in F9 embryonal carcinoma cells. Nucleic Acid Res 19: 6087-6092

Wang JC (1985) DNA topoisomerases. Annu Rev Biochem 54:665-697

Wood ER, Earnshaw WC (1990) Mitotic chromatin condensation in vitro using somatic cell extracts and nuclei with variable levels of endogenous topoisomerase II. J Cell Biol 111: 2839-2850

Dowsett CS, Clarke DJ, Mullinger AM, Gimenez-Abian JF, Creighton AM, Johnson RT (1994) A topoisomerase II-dependent G2 cycle checkpoint in mammalian cells. Nature 372: 467-470

Epstein FH, Smith PJ (1988) Estrogen-induced potentiation of DNA damage and cytotoxicity in human breast cancer cells treated with topoisomerase II-interactive antitumor drugs.

HEMOPOIETIC CELL DIFFERENTIATION AND DEATH BY RETINOIDS.

Domenico Delia, Antonella Aiello , Marco A Pierotti
Division OSA,
Istituto Nazionale Tumori
Via G Venezian 1
20133 Milano
Italy

The continous production of functionally mature cell populations depends on a complex series of tightly controlled proliferative, differentiative and survival events (Ogawa,1993). The more precise understanding of the biological and molecular mechanisms underlying these events provides valuable insights into normal and neoplastic cell transformation and should also suggest new directions for therapeutic intervention.

Retinoids play a central role in mammalian developmental and physiological functions as result of their profound effect on cell growth, differentiation, survival and death by apoptosis (Gudas,1992). Retinoids elicit biological responses by binding to and activating specific receptors, RARs and RXRs, members of the steroid/thyroid superfamily of nuclear molecules that act as transcription factors (Mangelsdorf,1994). It has been recently shown that some retinoids do not induce transcriptional activation, but rather transcriptional repression through an anti-AP1 activity (Fanjul et al,1994).

Activation and/or repression of gene transcription by retinoids profoundly modulates cell growth and differentiation and for such reason these compounds have considerable pharmacological interest (Hong and Itri,1994).

During the past decade it has been extensively demonstrated, using cell culture systems and animal models, that transformed cells can undergo terminal differentiation and loss of the neoplastic behaviour in response to all-trans retinoic acid (ATRA), as in the case of acute promyelocytic leukemia(APL). These findings have opened new prospects for the clinical use of retinoids in cancer differentiation therapy (Smith et al,1992). The recent results of the dramatic antitumor efficacy of ATRA in patients with APL (Degos,1994) represent a major clinical advance and has generated great expectancy.

The therapeutic use of retinoids, however, is often limited by their acute, systemic and chronic toxicity as well as teratogenicity (Armstrong et al,1994).

NATO ASI Series, Vol. H 95
Flow and Image Cytometry
Edited by Alain Jacquemin-Sablon

For this reason, several new retinoid derivatives have been developed, in the hope to identify those with retinoid activity but low toxicity.
We have used the HL-60 and NB4 cells as a model to characterize the effect of new clinically relevant retinoids on growth, differentiation and apoptosis. In particular, we have investigated the biological effects of the synthetic retinoid N-(4-hydroxyphenyl)all-trans-retinamide (HPR), whose efficacy in the chemoprevention of tumors in carcinogen-treated animals is well demonstrated (Moon et al,1994). In humans, HPR exhibits much lower toxicity and a more favorable long-term tolerability than ATRA (Formelli et al,1993). Currently HPR is in clinical trials in our Institute to determine its efficacy in preventing contralateral tumors in women surgically treated for node-negative breast cancer (Veronesi et al,1992; De Palo et al,1995) and in preventing relapses in patients surgically treated for oral leukoplakia (Chiesa et al,1992).
Over the last years we have thoroughly investigated and compared the biological and molecular responses of HPR to those of ATRA to elucidate its mechanism of action. We have found that HPR has not only anti-proliferative activity against several malignant hemopoietic cell lines, but also potent apoptotic activity (Delia et al,1993). These data have been recently confirmed by other investigators in cell lines of non-hemopoietic origin (Mariotti et al,1994).
Here we summarize the results obtained to data by comparing the effects of HPR and ATRA, and which suggest that these retinoids act through distinct pathways.

Differential antiproliferative effects of HPR and ATRA on leukemia and lymphoma cells:

When the effects of HPR and ATRA on the growth of human hemopoietic cell lines are compared, clear differencies emerge, as shown by the 72-hour dose-response analysis (Fig 1). It can be seen that the B-lymphoblastoid cell lines DHL-4, DoHH2, K422 and Raji, the former three bearing the t(14;18) chromosomal translocation and molecular rearrangements of the bcl-2 gene, are all growth-inhibited by HPR, though to a various degree. The effect, however, is more marked in DoHH2 cells. By contrast, the growth suppression caused by an equimolar dose of ATRA is less marked, except in the K422 cells which are extremely responsive to ATRA.
Although the growth of HL-60 cells is inhibited by both HPR and ATRA, the growth of HL-60R, a HL60 subclone with a point mutation in the RARa which confers resistace to ATRA, is inhibited by HPR only. Likewise, whereas NB4,

which carries the t(15;17) translocation typical of promyelocytic lekemias (Lanotte et al,1991), is sensitive to HPR and ATRA, the NB306 subclone, deficient in pml/RARa protein and thus resistant to ATRA (Dermine et al,1993), is responsive to HPR only. Altogether these data indicate that ATRA and HPR markedly differ in their antiproliferative activity. Moreover, the data with the RAR-alpha cell mutants indicate that ATRA and HPR may operate through distinct receptors and/or pathways.

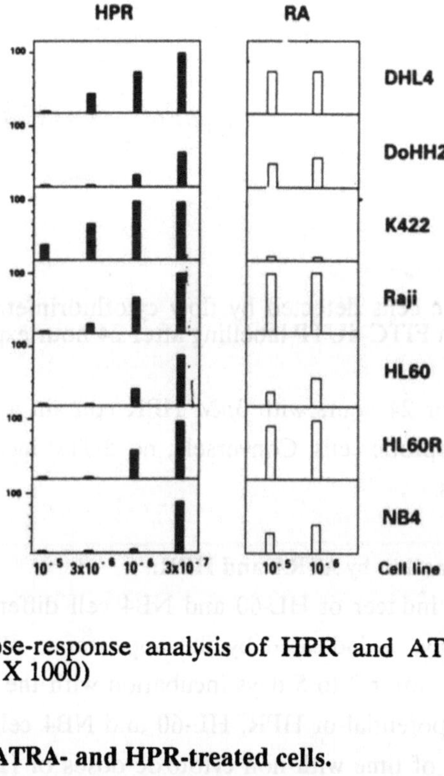

Figure 1: 72-hour dose-response analysis of HPR and ATRA. 3H-thymidine uptake results (values X 1000)

Cell cycle analysis of ATRA- and HPR-treated cells.

Flow cytometric DNA analysis of propidium iodide (PI) stained HL-60 cells show that treatment for 72 hours with 1 to 10μM ATRA results in a reduction in the percentage of S/G2M cells (Fig 2,left). By contrast, treatment with 0.3 to 3 μM causes a striking dose-dependent increase in the fraction of cells with a subdiploid DNA content (Fig 2,left). This finding, suggestive of cells undergoing death by apoptosis and containing fragmented nuclear DNA (Nicoletti et al,1991), has been further substantiated by the in-situ incorporation of fluorescent dUTP technique (Gavrieli et al,1992). It can be seen (Fig 2,right)

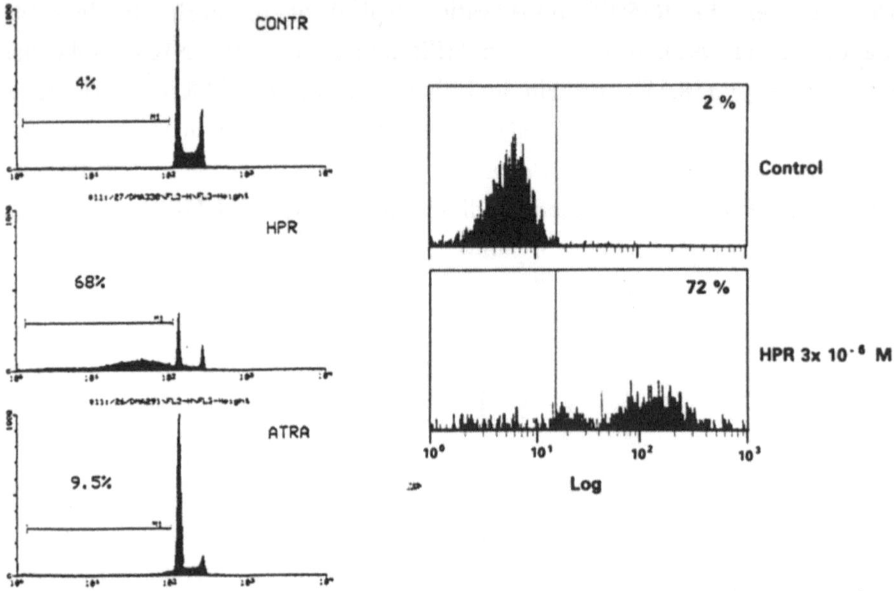

Figure 2: Apoptotic cells detected by flow cytofluorimetry on samples stained with PI(left) or with FITC-dUTP labelling after 24 hour exposure to HPR.

that cells treated for 24 hours with 3μM HPR contain a significant number of dUTP-labelled, apoptotic cells. Conversely, no dUTP incorporation is found in ATRA-treated cells.

Differentiation induction by ATRA and HPR:

ATRA is a potent inducer of HL-60 and NB4 cell differentiation towards the granulocytic pathway, as indicated by the appearance of cytoplasmic granules and indented nuclei after 2 to 5 days incubation with the retinoid. To establish the differentiative potential of HPR, HL-60 and NB4 cells have been cultured for various periods of time with non cytotoxic doses of HPR (below 1μM) and examined for morphology by May-Grunwald-Giemsa staining. Unlike ATRA, HPR is unable to promote in these cells granulocytic differentiation.

We have previously shown that differentiation of HL60 towards granulocytes or macrophage-like cell by DMSO or TPA, respectively, is associated with a progressive decline in bcl-2 gene expression (Delia et al,1992). The results of the treatment of HL60 cells with 1μM ATRA or HPR on bcl-2 expression are shown in Fig 3 . It can be seen that whereas the levels of bcl-2 protein decline upon treatment with ATRA, similar to what is observed following DMSO or TPA treatment (Delia et al,1992), remain unchanged after treatment with HPR. We

have also found that doses of HPR (>1μM) which induce apoptosis within 24 to 48 hours have no influence on bcl-2 protein expression (Delia et al,1995).
CD38 is a leukocyte antigen and bifunctional enzyme catalyzing the synthesis and hydrolysis of of cyclic ADP-ribose (Howard et al,1993). In HL60 cells this molecule is rapidly induced in a differentiation-independent manner by ATRA, but not by DMSO or TPA (Drach et al,1994). We have found that CD38 is also induced in cells treated for 24 hours with non-cytotoxic doses of HPR (Fig 3), though less strongly than ATRA. This result is rather unexpected when considering that ATRA and HPR responses are remarkably different, and suggest that these retinoids could share some signalling pathways.
Overall, these findings indicate that, unlike ATRA, HPR is unable to promote promyelocytic cell differentiation towards granulocytes. Moreover, that bcl-2 downregulation is related to the differentiation process and that apoptosis by HPR does not occur through direct modulation of the bcl-2 gene.

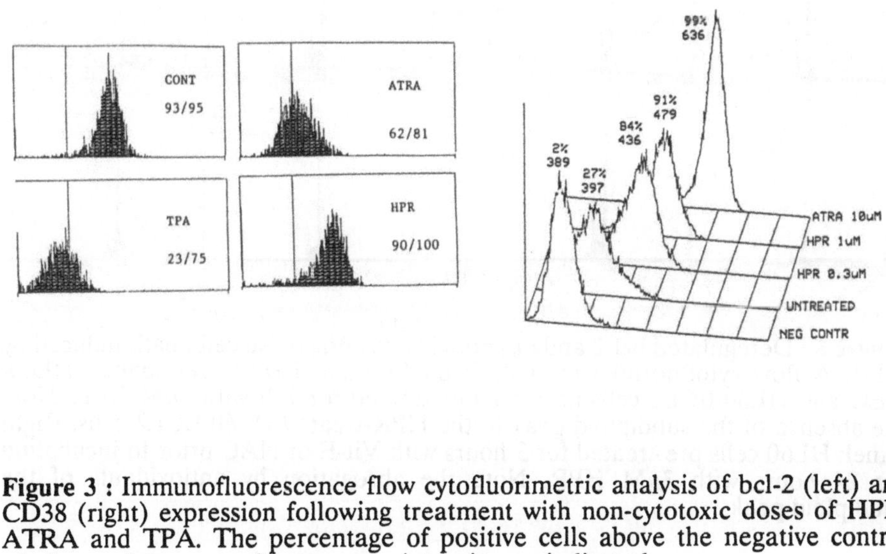

Figure 3 : Immunofluorescence flow cytofluorimetric analysis of bcl-2 (left) and CD38 (right) expression following treatment with non-cytotoxic doses of HPR, ATRA and TPA. The percentage of positive cells above the negative control window and the mean fluorescence intensity are indicated

HPR-induced cell death is antagonised by bcl-2 and antioxidants.

The bcl-2 gene is a repressor of apoptosis in a variety of cellular systems induced by growth factor withdrawal and by chemotherapeutic agents (Reed,1994; Korsmeyer, 1995). HL60 cells transfected with bcl-2 cDNA (HL60-BCL-2) (Naumovski and Cleary,1994) or with a control plasmid (HL60-NEO) have been analysed by flow cytofluorimetry to determine the degree of protection afforded by bcl-2 in response to HPR. It can be seen (Fig 4,left) that whereas HL60-NEO cells are to a large extent apoptotic after 72-hours of treatment with HPR, HL60-BCL-2 cells cultured under similar conditions are mostly viable. However, the viability of the latter cells declines with the incubation time, indicating that bcl-2 markedly delays, but does not abrogate the apoptosis induced by HPR. These findings are concordant with previous results (Delia et al,1995).

As bcl-2 has been shown to be involved in an antioxidative process (Kane et al,1993; Hockenbery et al,1993), we determined whether HPR-triggered apoptosis is prevented also by antioxidants. The results (Fig 4) indicate that non-cytotoxic doses of N-acetyl-l-cystein(NAC), Vitamin-C (Vit-C), Vitamin-E(Vit-E) and Deferoxamine suppress cell death induced by 3μM HPR. Hence, cell death by HPR is somehow associated with oxidative stress.

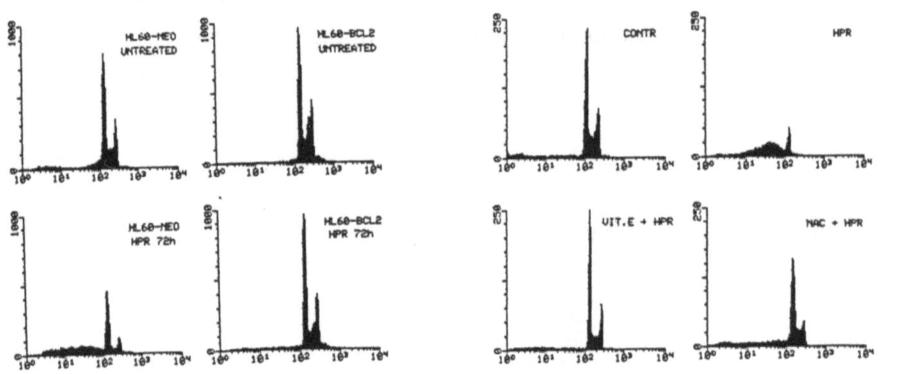

Figure 5 : Deregulated bcl-2 and of antioxidants antagonise cell death induced by HPR. A flow cytofluorimetric analysis on PI stained cells. Left panel: HL60-NEO and HL60-BCL2 cells untreated or treated for 72h with 3μM HPR. Note the absence of the subdiploid peak in the HPR-treated HL60-BCL2 cells. Right panel: HL60 cells pre-treated for 3 hours with Vit-E or NAC prior to incubation for 24 hours with 3μM HPR. Note the abrogation by antioxidants of the subdiploid peak.

Concluding remarks:

Several synthetic retinoids have marked antiproliferative and cytodifferentiative activity on a wide range of tumor cells (Gudas et al,1992; Kizaki et al,1993; Fanjul et al,1994). These features make retinoids important agents in cancer therapy, alternative to conventional chemotherapy, as in the case of acute promyelocytic leukemia where ATRA is now the primary agent for the therapy of this disease (Degos,1994; Grignani et al,1994). Retinoids are also candidates for chemoprevention since they inhibit several aspects of the carcinogenic process (transformation, metaplasia, promotion) (Hong and Itri,1994).

HPR is a synthetic retinoid derivative initially shown to inhibit the incidence of breast and prostate tumors in rats treated with MNU or DMBA (Meyskens et al,1983; Moon et al,1994). Such promising studies have recently been extended to humans, and two major random clinical trials are underway in our Institute to evaluate the efficacy of HPR in reducing the incidence of contralateral breast cancer in women (Veronesi et al,1992) and oral cavity tumors (Chiesa et al,1992).

Despite the clinical relevance of HPR, little is known about its molecular mode of action and only recently we have begun elucidating the biological responses to HPR. The accompaning data, along with those reported elsewhere (Delia et al,1993; Delia et al,1995), clearly indicate that HPR has antiproliferative activity, often more marked than equimolar doses of ATRA. Taking advantage of cell lines with RAR-alpha mutations, thus unable to respond to ATRA, we have demonstrated that these cells have a normal response to HPR, suggesting that HPR may not mediate its action via the classical retinoid receptor pathway. Concordant with this possibility is the evidence that HPR fails to bind significantly RAR proteins (Sani,1993) and to transactivate RAR or RXR (PG Pelicci, personal communication).

We have shown that non-cytotoxic doses of HPR induce CD38 antigen upregulation in HL60 cells, though less potently than ATRA. Induction of this molecule represents a specific response to ATRA, rather than a general response to differentiation, and is mediated by RAR-alpha in HL60 cells (Drach et al,1994). The upregulation of CD38 by HPR is rather unexpected given the evidence that HPR responses appear not to involve RAR or RXR. The significance of this finding remains unclear.

It has been recently found that certain retinoids, despite being unable to induce differentiation, effectively inhibit proliferation of cancer cells through an anti-AP1 activity (Fanjul et al,1994). Retinoids with these features are of clinical interest as they are thought to have reduced side effects. Although HPR shares similar biological activities, it remains unknown whether this retinoid exhibits an anti-AP1 activity, responsible for the antiproliferative effect.

We have demonstrated that pharmacologically relevant doses of HPR (1 to 3μM) induce apoptotic cell death. Apoptosis has also been observed in ATRA-treated cells, however, in this case the phenomenon is secondary to terminal differentiation and is associated with downregulation of the anti-apoptotic gene bcl-2 (Delia et al,1992). By contrast, apoptosis by HPR appears to be a primary event, and does not involve bcl-2 modulation.

Bcl-2 is a gene whose expression protects from various, but not all forms of cell death (Reed,1994; Korsmeyer, 1995). We have demonstrated that deregulated expression of bcl-2 by HL60 cells significantly prolongs their survival in response to 3μM HPR. This finding, coupled with recent data indicating that bcl-2 may be involved in an antioxidative pathway (Hockenbery et al,1993; Kane et al,1993) and that oxidative stress is associated with apoptosis (Buttke and Sandstrom,1994), has lead us to verify whether also antioxidants provide protection against HPR. The results clearly show that water soluble (Vit-C, NAC,GSH) and lipid-soluble (Vit-E) antioxidants antagonise the effect of HPR and suggest that this retinoid elicits the formation of intracellular oxidants which eventually trigger apoptosis.

We believe that the results thus far obtained may have a number of clinical implications. The demonstration that HPR elicits apoptotic cell death at pharmacologically achievable doses in humans, highlights the potential role of HPR for an apoptosis-based therapy of cancer patients. Moreover, the evidence that ATRA and HPR do not confer cellular cross-resistance, suggests that HPR could find an application in the treatment of APL that have become resistant to ATRA (Delva et al,1993). Finally, the data raise a major, and unanswered question as to whether oxidative stress and apoptosis are the relevant players in cancer chemoprevention.

Acknowledgments:

This work was partially supported by the Associaziona Italiana Ricerca Cancro (AIRC) and by the CNR Target Project PF-BTBS. We wish to thank Dr Naumovski and M.Cleary for providing the HL60-BCL2 cells.

References:

Armstrong RB,Ashenfelter KO,Eckhoff C,Levin AA,Shapiro SS. General and reproductive toxicology of retinoids. In The Retinoids: biology,chemistry and medicine. 2nd Edition. Sporn MB and Roberts AB and Goodman DS eds, Raven Press, Ltd,New York,1994, pg 547-572.

Buttke TM,Sandstrom PA. Oxidative stress as a mediator of apoptosis. Immunol Today,15:7-10,1994.

Chiesa F, Tradati N, Marazza M, Rossi N, Boracchi P, Mariani L, Clerici M, Formelli F, Barzan L, Carrassi A, Pastorini A, Camerini T, Giardini R, Zurrida S, Minn FL, Costa A, De Palo G, Veronesi U: Prevention of local relapses and new localizations of oral leukoplakias with the synthetic retinoid fenretinide (4-HPR). Preliminary results. Oral Oncol, Eur J Cancer, 28B, 97-102, 1992.

Degos L. Is acute promyelocytic leukemia a curable disease? treatment strategy for a long term survival. Leukemia, 8:911-913,1994.

Delia D, Aiello A, Soligo D, Fontanella E, Melani C, Pezzella F, Pierotti MA, Della Porta G. Bcl-2 proto-oncogene expression in normal and neoplastic human myeloid cells. Blood, 79, 1291-1298, 1992.

Delia D, Aiello A, Lombardi L, Pelicci PG, Grignani FJr, Grignani F, Formelli F, Menard S, Costa A, Veronesi U, Pierotti MA: N-(4-hydroxyphenyl)retinamide induces apoptosis of malignant hemopoietic cell lines including those unresponsive to retinoic acid. Cancer Res, 53:6036-6041, 1993.

Delia D, Aiello A,Formelli F, Fontanella E,Costa A, Miyashita T,Reed JC,Pierotti MA. Regulation of apoptosis induced by the retinoid N-(4-hydroxyphenyl) retinamide and effect of deregulated bcl-2. Blood,85:359-367,1995.

Delva L, Cornic M, Balitrand N, Guidez F, Miclea JM, Delmer A, Teillet F, Fenaux P, Castaigne S, Degos L, Chomienne C. Resistance to all-trans retinoic acid (ATRA) therapy in relapsing acute promyelocytic leukemia: study of in vitro ATRA sensitivity and cellular retinoic acid binding protein levels in leukemic cells. Blood, 82:2175-2181, 1993.

De Palo G,Veronesi U,Camerini T,Formelli F,Mascotti G,Boni C,Del Vecchio M,Campa T,Costa A,Marubini E. Can Fenretinide protect women against ovarian cancer?. JNCI,87:146-147,1995.

Dermine S,Grignani F,Clerici M,Nervi C,Sozzi G,Talamo GP,Marchesi E, Formelli F,Parmiani G,Pelicci PG, Gambacorti-Passerini C. The occurrence of resistance to retinoic acid in the acute promyelocytic leukemia cell line NB306 is associated with altered expression of pml/RAR protein. Blood,82:1573-1577,1993.

Drach J,McQueen TM,Engel H,Andreff M,Robertson KA,Collins SJ,Malavasi F,Metha K. Retinoic acid-induced expression of CD38 antigen in myeloid cells is mediated trough the retinoic acid receptor-alpha. Cancer Res,54:1746-1752,1994.

Fanjul A,Dawson MI,Hobbs PD,Jong L,Cameron JF,Harlev E,Graupner G,Lu XP,Pfahl M. A new class of retinoids with selective inhibition of AP1 inhibits proliferation. Nature,372:107-11,1994.

Formelli F, Clerici M,Campa T,Di Mauro MG,Magni A,Mascotti A,Moglia D,De Palo G,Costa A,Veronesi U. Five year administration of fenretinide:pharmacokinetics and effects of plasma on retinol concentrations. Cancer Res,11:2036-2042,1993.

Gavrieli Y, Sherman Y, Ben-Sasson SA. Identification of programmed cell death in situ via specific labeling of nuclear DNA fragmentation. J Cell Biol,119:493-501,1992.

Grignani F, Fagioli M,Alcalay M,Longo L,Pandolfi PP,Donti E,Biondi A,Lo Coco F,Grignani F,Pelicci PG. Acute promyelocytic leukemia: form genetics to treatment. Blood,83:10-25,1994.

Gudas JL. Retinoids,retinoid-responsive genes,cell differentiation and cancer. Cell growth Diff,3:365-371,1992.

Hockenbery DM, Oltvai ZN, Yin XM, Milliman CL, Korsmeyer SJ: Bcl-2 functions in an antioxidant pathway to prevent apoptosis. Cell, 75:241-251, 1993.

Hong WK and Itri LM. Retinoids and human cancer. In In The retinoids :biology,chemistry, and medicine. Eds Sporn MB,Roberts AB,Goodman DS. pg597-630,1994. Raven Press,NY

Howard M,Grimaldi JC,Bazan JF,Lund FE,Santos-Argumendo L,Parhouse RM,Walseth TF,Lee HC. Formation and hydrolysis of cyclic ADP-ribose catalyzed by lymphocyte antigen CD38. Science,262:1056-1059,1993.

Kane DJ, Sarafian TA, Anton R, Ha H, Gralla EB, Valentine JS, Ord T, Bredesen DE. Bcl-2 inhibition of neural death:decreased generation of reactive oxygen species. Science, 262:1274-1277, 1993.

Kizaki M, Ikeda Y, Tanosaki R, Nakajima H, Morikawa M, Sakashita A, Koeffler HP. Effects of novel retinoic acid compound, 9-cis-retinoic acid, on proliferation, differentiation, and expression of retinoic acid receptor-Á and retinoid X receptor-Á RNA by HL-60 cells. Blood, 82:3592-3599,1993.

Korsmeyer SJ. Regulators of cell death. Trends in Genetics,11:101-105,1995.

Lanotte M, Martin-Thouvenin V, Najiman S, Ballerini P, Valensi F, Bergen R. NB4, a maturation inducible cell line with t(15;17)marker isolated from a human acute promyelocytic leukemia(M3). Blood, 77:1080-1086, 1991.

Lehmann JM, Jong L, Fanjul A, Cameron JF, Lu XP, Haefner P,Dawson MI, Pfahl M. Retinoids selective for retinoid X receptor response pathways. Science,258:1944-1946,1992.

Moon RC, Metha RG, Rao KVN. Retinoids and cancer in experimental animals. In The Retinoids:biology,chemistry and medicine. 2nd Edition. Sporn MB and Roberts AB and Goodman DS eds, Raven Press,Ltd,New York,1994, pg 573-595.

Mangelsdorf DJ,Umesono K,Evans RM. The retinoid receptors. In The retinoids:biology,chemistry and medicine. Eds Sporn MB,Roberts AB,Goodman DS. pg319-350,1994. Raven Press,NY.

Mariotti A, Marcora E, Bunone G, Costa A,Veronesi U, Pierotti MA, Della Valle G. N-(4-hydroxyphenyl) retinamide: a potent inducer of apoptosis in human neuroblastoma cells. J Natl Cancer Inst,86:1245-1247,1994.

Meyskens FL, Alberts DS, Salmon SE. Effect of 13-cis-retinoic acid and 4-hydroxyphenyl-all-trans-retinamide on human tumor colony formation in soft agar. Int J Cancer, 32:295-299, 1983.

Naumovski L,Cleary ML. bcl-2 inhibits apoptosis associated with terminal differentiation of HL-60 myeloid leukemia cells. Blood,83:2261,1994.

Nicoletti I, Migliorati G,Pagliacci MC,Grignani F,Riccardi C. A rapid and simple method for measuring thymocyte apoptosis by propidium iodide staining and flow cytometry. J Immunol Methods,139:271,1991.

Ogawa M. Differentiation and proliferation of hemopoietic stem cells. Blood,81:2844-2853,1993.

Reed JC. Bcl-2 and the regulation of programmed cell death. J Cell Biol, 124:1,1994.

Sani BP. Binding affinities of retinoids to their binding proteins/receptors and their biological activities.In Retinoids:progress in research and clinical applications.Eds MA Livrea and Packer L. Marcel Dekker Inc,NY,1993,pg 337-247.

Smith MA,Parkinson DR,Cheson BD,Friedman MA. Retinoids in cancer therapy. J Clin Oncol,10:839-864,1992.

Veronesi U, De Palo G, Costa A, Formelli F, Marubini E, Del Vecchio M. Chemoprevention of breast cancer with retinoids. J Natl Cancer Inst Monographs, 12:93-97, 1992.

PART III

Data Management Systems
Cell Sorting Techniques
Microscopy

Myc in the Control of Proliferation and Apoptosis

Lamorna Brown and Gerard Evan
Biochemistry of the Cell Nucleus
Imperial Cancer Research Fund
44 Lincoln's Inn Field
London
WC1A 3PX

Introduction

The c-myc gene is the cellular homologue of the viral oncogene v-myc, found in a number of avian retroviruses. Since its discovery, c-myc has been implicated in the development of human cancers and it has been found to be altered in a variety of different tumours. Alterations include duplication of the gene and de-regulated expression through a number of mechanisms, for example, activation due to chromosome translocation.

c-myc has been implicated in the control of normal cellular proliferation. It is one of the immediate early growth response genes, and it is rapidly induced in quiescent cells upon mitogenic stimulation, suggesting that it may a role in the transition from quiescence to proliferation. Unlike many other early response genes which peak within hours of stimulation and then are shut off, Myc expression is maintained at a low level throughout the cell cycle and it is only down regulated on mitogen withdrawal.

The product of the c-myc gene, Myc, possesses a number of functional domains found in transcription factors. The carboxy-terminal helix-loop-helix motif and leucine-zipper motif are found in other proteins which regulate transcription and are thought to act as dimerisation domains between proteins. Indeed, Myc has been shown to dimerize to

NATO ASI Series, Vol. H 95
Flow and Image Cytometry
Edited by Alain Jacquemin-Sablon

Max via the helix-loop-helix domain. The adjacent basic DNA-binding domain recognises an E-box enhancer motif. In addition, Myc has a putative trans-activation domain at the amino-terminus. Analysis of the action of these domains indicates that Myc probably functions as a transcription factor, however the targets of Myc regulation remain elusive.

Expression of Myc in growth factor deprived fibroblasts is sufficient to drive cells into proliferation. Continued expression of Myc prevents cells from undergoing differentiation programs and can block growth arrest. These observations suggest that Myc regulates expression of genes that execute all of these processes. Furthermore it is possible that down regulation of Myc expression is required, or is a signal, for growth arrest. If this is the case, continued expression of Myc would prevent growth arrest and lead to continued proliferation, as seen in tumour cells.

De-regulated expression of c-Myc in Rat1 fibroblasts

We wished to determine the effects of de-regulated expression of Myc in Rat1 fibroblasts. It was predicted that constitutive expression of c-myc, which could not be down regulated upon mitogen withdrawal, would prevent the normal growth arrest imposed by serum starvation. Surprisingly, the growth curve of cells expressing constitutively expressed Myc did not differ significantly from the control cells. However, analysis of the DNA content and BrdU incorporation of the cells indicated that the cells constitutively expressing Myc failed to go into growth arrest, although there was no increase in overall cell number. The obvious reason for this would be that the Rat1/Myc cells were dying as fast as they were proliferating.

There are two significantly different ways in which a cell can die. The firsts and longest recognised way is by necrosis. In this event the cell swells and bursts, internal organelles are damaged and the contents of the cell are released into the local environment, often leading to inflammation. Necrosis is not an active process of the cell and cannot be control by cell autonomous genetic events. More recently a second form of cell death has been recognised. This is termed apoptosis or programmed cell death. It is an active process which can be controlled by genetic events within an individual cell. It leads to condensation of the nucleus, often accompanied by degradation of the DNA into a nucleosome ladder. The cell shrinks and blebs giving rise to apoptotic vesicles which contain the contents of the cell. The remnants of the cells are phagocytosed by neighbouring cells, thereby preventing inflammation.

In our experiments a culture of Rat1/Myc cells was examined by time-lapse video microscopy. This confirmed that the cells were indeed dying and indicated that the deaths were by apoptosis. The cells underwent the classic stages of apoptosis, including cytoplasmic shrinkage, nuclear condensation and vesiculization. This was accompanied by appearance on a nucleosome ladder.

The regions of Myc that were necessary for induction of apoptosis in this system were analysed using mutants of Myc. From these studies it was apparent that the regions required are co-incident with the regions of Myc that are necessary for co-transformation and inhibition of differentiation. These regions include the basic-helix-loop-helix-leucine-zipper region and the transactivation domain at the amino terminus. Furthermore, a series of elegant experiments carried out by Bruno Armati and Trevor Littlewood, demonstrated that Myc must interact with is dimerisation partner Max, in order to activate the apoptotic pathway. This suggests that Myc induces apoptosis through its activity as a transcription factor. However, in an apparent contradiction to this, cycloheximide treatment of Myc expressing cell did not prevent apoptosis, suggesting that *de novo* protein synthesis is not required. This implies that expression of Myc establishes the apoptotic pathway, which is then held in check by some mechanism until condition require the execution of the apoptotic process.

Why does Myc induce apoptosis?

The above information indicates that although Myc expression usually leads to proliferation, when the cell cycle is blocked by removal of serum apoptosis is also induced. This gave rise to two models of Myc action. The first is termed the CONFLICT OF SIGNALS model. This model presumes that the major role of Myc is to drive proliferation. When Myc is expressed in an inappropriate environment the cell does not receive the necessary mitogenic signals from growth factors and a conflict of signals is perceived - proliferating signals from Myc versus arrest signals from the environment. This results in a crisis for the cell which does not know which path to follow and therefore undergoes

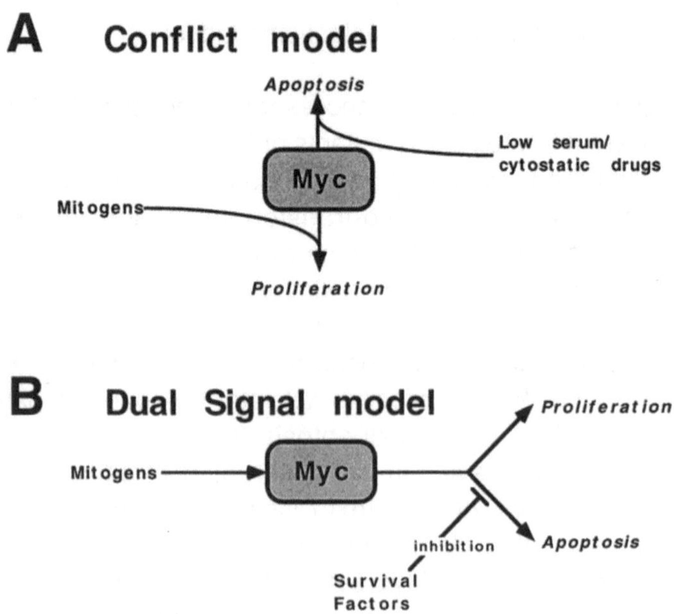

Figure 1. A representation of the two models of Myc action in inducing apoptosis. A. The Conflict of Signals Model. Myc induces either proliferation or apoptosis, depending on the availability of secondary mitogenic signals. B. The dual Signal Model. Myc induces both proliferation and apoptosis but the apoptotic pathway must be blocked to allow proliferation.

apoptosis - since it does not know what else to do! In this model, apoptosis is not a normal function of Myc but rather a pathological result of its inappropriate expression.

The second model, which we favour, is the DUAL KEY MODEL. In this model, expression of Myc sets up two opposing pathways, one for proliferation and one for apoptosis. The cell must receive additional signal to suppress apoptosis in order to go down the proliferation pathway. If this second signal is not received, the apoptotic events are executed instead. This model suggests that both proliferation and apoptosis are normal functions of Myc. Indeed, it would seem prudent that a pathway that could so easily drive uncontrolled proliferation should have a counter-check to suppress it.

One of the key differences between the two models is the nature of the second signal. The "Conflict" model requires that the second signal is mitogenic and reinforces the Myc signal. In contrast, the "Dual Key" model predicts that the second signal will be anti-apoptotic but need not be mitogenic.

<u>IGF-I suppresses Myc induced apoptosis</u>

As already stated, Myc induces apoptosis in the absence of serum. Presumably components of serum suppress Myc induced apoptosis and initial experiments indicated that it was the absence of defined growth factors rather that the lack of metabolites that promoted Myc induced cell death. We examined the effect of adding back a series of defined growth factors to serum starved Rat I cells containing over expressed Myc.

In these experiments we used an inducible Myc protein, MycER. This is a chimera of the entire Myc protein fused to the oestrogen receptor. In the absence of oestrogen the chimeric protein adopts a confirmation in which Myc is inactive, however when oestrogen is added and binds to its receptor the confirmation is altered and Myc becomes active. Thus these cells could be serum starved without induction of apoptosis. Subsequent addition of β-oestradiol activates Myc and induces apoptosis as seen previously. However, if serum is replaced by certain growth factors, activation of Myc did not induce apoptosis. These growth factors include IGF-I IGF-II, PDGF and insulin. Other growth factors such as EGF, FGF and bombesin, did not suppress Myc induced apoptosis even at levels 10 fold higher than those at which these growth factors are used mitogenically. Immediately this suggests that mitogenic activity of a cytokine is not sufficient to prevent apoptosis - in contradiction to the conflict model of Myc induced apoptosis.

Drugs that arrest cells can also be used to trigger Myc induced apoptosis in place of serum starvation. Thymidine is a DNA synthesis inhibitor and blocks cells in S phase, while etoposide is a topoisomerase II inhibitor and block cells in S/G2. Both drugs block after the G1 commitment point beyond which the cells no longer depend on mitogenic growth factors for cell cycle progression. After drug treatment, Rat1/MycER cells were treated with β-oestradiol and analysed by time-lapse in the presence or absence of IGF-I. Addition of IGF-I significantly delays and suppresses apoptosis in the presence of either drug - conditions in which IGF-I could not be acting as a mitogen. Hence it is unlikely that the anti-apoptotic effect of IGF-I is linked to its mitogenic activity.

Co-operation of oncogenes in tumour progression

The observation that Myc induces cell death when proliferation is blocked, is at odds with the notion that Myc is involved in the tumour progression . If this were the case, it would be expected that tumours with aberrantly expressed Myc, that out grow the local supply of growth factors, would spontaneously regress due to the Myc induced apoptosis. Clearly this does not happen so the survival of tumours with de-regulated expression of Myc is a paradox unless it is explained by some abnormality of the apoptotic pathway.

Over expression the Bcl2 gene has been demonstrated to prevent apoptosis. Bcl2 was first identified as the gene at the translocation breakpoint in B-cell Lymphoma. As a result of the 14:18 chromosome translocation the Bcl2 gene is juxtaposed with the Ig heavy chain gene and is over expressed, presumably contributing to the generation of the lymphoma. The Bcl2 gene product is a 26 kDa protein that has been shown to inhibit apoptosis in model systems, although its mechanism of action is unknown. A novel form of oncogene co-operation may occur if over expression of Bcl2 in tumours is sufficient to ensure growth in the presence of de-regulated Myc. In our experiments the over expression of Bcl2 in the Rat1/MycER cell was found to suppress the

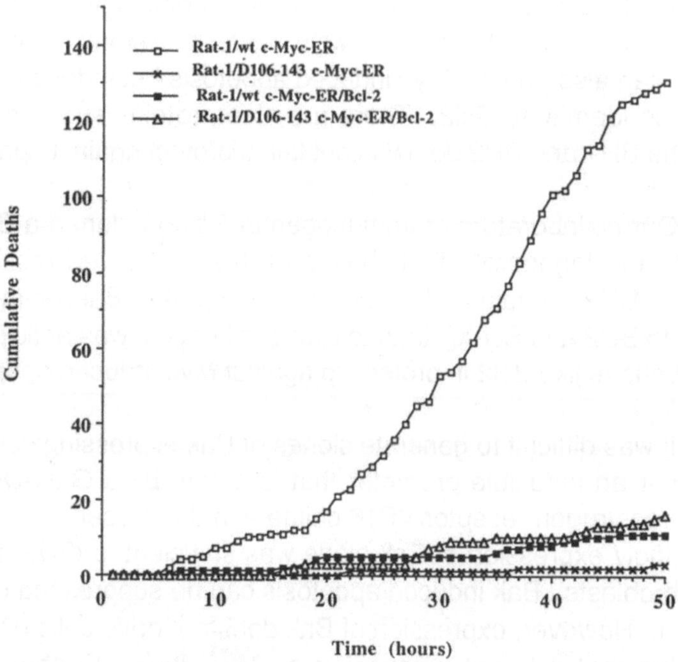

Figure 2. Time, lapse videomicroscopic quantification of apoptosis in Rat-1 cells. Rat-1 cells expressing either wild-type c-MycER or the defective c-Myc mutant Δ106-143 fused to the human oestrogen receptor, and either with or without Bcl-2, were growth arrested by culture in 0.5% serum for 48 hours. Activation of Myc was acomplished by additionof β-oestradiol and the cells were observed by time lapse video microscopy. Images were taken at a rate of 12 frames per hour. Apoptotic cell deaths were scored and plotted as cumlative deaths against time.

induction of apoptosis on removal of serum (Fig 2). Myc induced apoptosis can therefore be suppressed by factors such as IGF-I or by expression of genes such as Bcl2.

The role of Bcl2 family members in apoptosis

Over the past few years a number of Bcl2 related genes have been identified. Some of these share homology domains with Bcl2 while others are functionally homologous in that they prevent apoptosis in a variety of model systems although they lack amino acid identity to Bcl2. For example, Bcl-X_L share the BH1 and BH2 homology domains and Bcl-X_L can prevent apoptosis in the Myc induced system. The viral gene, $p19^{E1B}$, can also prevent Myc induced apoptosis but it does not have any sequence identity to Bcl2. There are also proteins such as Bax which share the BH1 and BH2 domains but fail to protect against apoptosis.

Our collaborators at Immunogen in Boston cloned a Bcl2 related protein, by degenerate PCR based on the conserved residues in the BH1 and BH2 domains. This protein, designated Bak, is more closely related to Bcl2 and Bcl-X_L than to Bax and hence it was anticipated that it would behave like Bcl2 in protecting against Myc-induced apoptosis.

It was difficult to generate clones of Bak expressing cells so it was put under an inducible promoter that is driven by a Gal4-DNA-binding domain/oestrogen receptor/VP16 chimera in the presence of Tamoxifen. Surprisingly expression of Bak alone was sufficient to drive apoptosis in Rat1 fibroblasts. Bak induced apoptosis can be suppressed by serum or by IGF-I. However, expression of Bak does not drive cells into cell cycle indicating that it is not functioning as Myc does. Further more, Bak antagonises the protection afforded by Bcl2 to Myc induced apoptosis in Rat1/MycER cells.

It is now clear that the Bcl2 family of proteins fall into two distinct groups, those that inhibit apoptosis and those that promote it. Recent work indicates that several of the family members interact through the BH1 and BH2 domains, possibly forming a complicated network. It seem likely that it is a balance between these positive and negative regulators that determines the fate of the cell, whether it live or dies. However, Myc and the Bcl2 family of proteins appear to be regulator of commitment to apoptosis rather than effectors of apoptosis. At present, the most likely effectors are the ICE-like proteases.

Conclusion

In summary, Myc is required for proliferation but the potential of cells to proliferate is controlled by a parallel pathway of apoptosis. If Myc is de-regulated in the absence of survival factors the apoptotic pathway is executed. IGF-I is a survival factor for fibroblasts, a property which is unlinked to its mitogenic activity. Bcl2 can also block Myc induced apoptosis but this function can be antagonised by the related protein Bak. Bak alone can induce apoptosis, which can be suppressed by IGF-I. This indicates that the role of IGF-I is down stream of the Bcl2 and Bak interaction. Nevertheless, the mechanism of IGF-I, Myc or the Bcl2 proteins in control of apoptosis is unclear at present.

References

Amati, B., Brooks, M., Levy, N., Littlewood, T., Evan, G. and Land, H. (1993). Oncogenic activity of the c-Myc protein requires dimerisation with Max. Cell., *72*, 233-245.

Amati, B., Littlewood, T., Evan, G. and Land, H. (1994). The c-Myc protein induces cell cycle progression and apoptosis through dimerisation with Max. EMBO J., *12*, 5083-5087.

Chittenden, T., Harrington, E., O'Connor, R., Evan, G. and Guild, B. (1995). Induction of apoptosis by the Bcl-2 homologue Bak. Nature., *374*, 733-736.

Fanidi, A., Harrington, E. and Evan, G. (1992). Cooperative interaction between *c-myc* and *bcl-2* proto-oncogenes. Nature., *359*, 554-556.

Harrington, E., Fanidi, A., Bennett, M. and Evan, G. (1994). Modulation of Myc-induced apoptosis by specific cytokines. EMBO J., *13*, 3286-3295.

Yang, E., Zha, J. P., Jockel, J., Boise, L. H., Thompson, C. B. and Korsmeyer, S. J. (1995). Bad, a heterodimeric partner for Bcl-xL and Bcl-2, displaces Bax and. Cell., *80*, 285-291.

Flow Cytometry: Analyses for All Sizes

James H. Jett
National Flow Cytometry Resource
Life Sciences Division
Los Alamos National Laboratory
Los Alamos, NM 87545 USA

In this article we will describe the breadth of application of flow cytometric analytical techniques in terms of size and complexity of the systems analyzed. As the current limits of application of flow cytometry are established, we will forecast areas in which the technology will have an impact in the future. Only a few illustrative examples will be cited here. An excellent, up to date, and complete set of references is contained in the book Practical Flow Cytometry, Third Edition (Shapiro 1994).

Table 1 presents a measurement scale that spans eight orders of magnitude along with examples of biological "particles" that have been analyzed at numerous points along the scale. The top of the scale, which is in microns, represents ten millimeters and the bottom of the scale represents 10 Angstroms. As Table 1 demonstrates, biological entities over the whole range have been analyzed by flow cytometry.

Initially, development of flow cytometric technology was focused on solving problems in mammalian cell biology. Thus the particles analyzed were simple, if a mammalian cell can be considered to be simple, and were typically in the range of from 5 to 10 microns in diameter. The history of the rapid rise of application of flow cytometry to quantification of properties of mammalian cells does not have to be repeated here. Suffice it to say that from the humble beginnings of measuring cellular DNA content, the labeling and measurement technology grew rapidly resulting in an tool that can be used to measure a wide variety of cellular properties.

NATO ASI Series, Vol. H 95
Flow and Image Cytometry
Edited by Alain Jacquemin-Sablon

Expansion of the technology to measurements of both larger and smaller particles took place simultaneously. While there are larger mammalian cells to be analyzed, initially the major driver of the technology toward the measurement of larger particles came from the plant science community. Plant cells, more particularly plant protoplasts (cells that have been stripped of their cellulose shell), are typically larger than mammalian cells. They range in size from 20 to 60 microns in diameter. Although many flow cytometers could analyze such particles, sorting was difficult to impossible with a typical 60 micron diameter orifice. The ability to sort larger particles by deflection of charged droplets was demonstrated to select plant protoplasts(Harkins and W. 1987), to purify fused plant protoplasts(Alexander, Cocking et al. 1985) and to size select multicellular spheroid models of tumors(Jett and Alexander 1985; Freyer, Wilder et al. 1987). The largest orifice used for droplet-based sorting is on the order of 400 microns(Davis, McAllister et al. 1988). A conservative estimate of the largest particle that can be conveniently sorted is half of the orifice diameter or 200 microns.

Gel microdroplets provide a second example of large particles analyzed and sorted by flow cytometry. This is a technique for isolating individual cells, eukaryotic or bacterial, originally pioneered by Weaver, et al(Weaver, Seissler et al. 1984; Weaver, Williams et al. 1988). Cells are isolated by encapsulating them in small (10-100 microns in diameter) droplets of agarose gel by a number of techniques such as acoustic gel droplet generation. During droplet generation, the average occupancy of the droplets by cells is kept low, on the order of 0.1, which results in only a small fraction of the droplets containing more than one cell(Nir, Lamed et al. 1990).

Analysis of the cellular content of each microdroplet can be used as a bacterial growth assay to determine antibiotic susceptibility(Sahar, Nir et al. 1994). Sorting of microdroplets provides a means of clonal selection of antibiotic resistant bacteria. The droplets can be considered as microenvironments surrounding individual cells which contain molecule shed by the encapsulated cell. The contents of these microenvironments can be assayed for the presence of specific molecules by antibody techniques.

Moving further up the size scale into the range above 100 microns, a variety of phytoplankton have been analyzed by flow cytometry. A phytoplankton analysis system has been developed for the study fresh water plankton colonies which range in diameter to 500 microns and to 2,000 microns in length(Dubelaar, Groenewegen et al. 1989; Peeters, Dubelaar et al. 1989). Recently, the Martinsreid group(Huller, Grossner et al. 1994) has described a new cytometer for the analysis of macroplankton in the 100-2,000 micron range. This instrument has several unique features including volume measurements based upon the Coulter principal and high resolution image acquisition. A new apparatus is under development for measuring ichthyoplankton that range is size up to 30,000 microns or 3 centimeters.

Turning to the bottom half of the size scale, one starts with particles that are in the micron size range. Mammalian chromosomes are most often analyzed, and sorted biological particles in this size range(Cram 1990; Gray and Cram 1990). Analysis techniques have become quite sophisticated with new techniques being developed to determine centromeric indices on commercial instruments(Rens, Van Oven et al. 1994). Future developments will make measurement of chromosome DNA content along with detection of one or more hybridization signals routine.

Bacterial genomes are approximately one thousandth the size of mammalian genomes and one tenth the diameter of mammalian cells which define the minimum instrument sensitivity required to analyze bacteria Numerous bacterial analyses have been performed by flow cytometry in a variety of specialized studies. Measurement of the AT/GC ratios in bacteria have been used as a means of species identification(Van Dilla, Langlois et al. 1983). AT/GC content analysis has also been used to measure the growth of bacteriophage T4 inside *Escherichia coli* cells prior to cell lysis(Sanders, Yajko et al. 1991). The measurements were robust enough to determine the GC content of the T4 bacteriophage to with an accuracy of 0.2% in the presence of the host *E. coli* DNA which has a different AT/GC ratio.

A high numerical aperture system has been developed(Steen 1990) which provides sensitivity for light scatter detection down to particles 0.2 microns in diameter and sufficiently high resolution DNA content measurements to observe integral chromosome numbers in *E. coli*. The methods developed have been applied to the studies of bacterial growth and antibiotic susceptibility.

Flow cytometry has been used to characterize directly, that is without sample culturing, organisms contained in water and fecal sources. Environmental water samples, concentrated by flocculation, were analyzed for the presence of *Cryptosporidium* and *Giardia* protozoa(Vesey, Hutton et al. 1994) by flow cytometry using immunofluorescence techniques. Direct analysis of anaerobic bacteria in fecal samples by detection of nucleic acids in the appropriate sized particles(van der Waaij, Mesander et al. 1994) has been combined with immunofluorescence detection of IgA-coated bacteria as a potential means of monitoring changes in gut flora.

In the submicron size range, two new species of marine organisms have been discovered by flow cytometry. A chlorophyll producing bacterium less than a micron in diameter was found (Chisholm, Olson et al. 1988)by combining forward angle light scatter measurements with Hoechst 33342 and chlorophyll fluorescence determinations. In other work, the smallest known eukaryotic organism was discovered (Courties, Vaquer et al. 1994)in marine algal samples.

Direct detection virus particles by light scatter measurements was reported a number of years ago(Hercher, Mueller et al. 1979). In a system tuned to detect orthogonal light scatter from small particles, a distinct light scatter distribution from T2 bacteriophages was recorded. Since this early work, there has not been much effort placed on direct analysis of virus particles.

Subcellular organelles, liberated from cells, have been analyzed by flow cytometry. Enzyme activity has been measured in membrane vesicles identified with immunofluorescence(Gorvel, Mishal et al. 1988). The existence of multiple endocytic vesicles in fluid phase endocytosis has also been demonstrated with flow cytometric techniques(Murphy 1985). Future applications will include the analysis of small liposomes for specific lipoproteins.

Recently, the sensitivity for fluorescence detection has been increased to the point that it is possible to detect individual molecules they pass through the laser beam (Nguyen, Keller et al. 1987; Shera, Seitzinger et al. 1991). One application of the increased sensitivity is the quantification of the size of DNA fragments. The increased sensitivity for fluorescence detection required for DNA fragment size measurements(Goodwin, Johnson et al. 1993) is achieved by decreasing the sample flow velocity such that the fragments are in the laser beam for several milliseconds and by resorting to photon counting. A bright, stoichiometrically binding dye such as TOTO(Glazer and Rye 1992) is used to stain the fragments. The fluorescence of TOTO is enhanced many-fold upon binding to DNA allowing the analysis of the fragments in the dye solution. The currently reported analysis limits(Petty, Johnson et al. 1995) are from 1.5 to 50 kilobases. Efforts are underway to extend the range even further in both directions. Ultimately, it should be possible to analyze the sizes of fragments down to the hundreds of bases and up into the megabase range.

This flow cytometric measurement has several advantages over conventional electrophoretic techniques used to measure DNA fragment size. First, it is fast. Even at low, by conventional flow cytometric standards, analysis rates of 50 or less fragments per second, it is possible to collect a data set in three minutes that allows size determinations to be made with an accuracy of better than two percent Since the amount of dye bound to the DNA fragment is proportional to its length, the amplitude of the signal produced is linearly proportional to the size of the DNA fragment. This is in sharp contrast to gel electrophoresis where the migration distance is inversely proportional the log of the molecular weight of the fragment. Often, in these measurements, the resolution achieved is determined by the counting statistics, that is, the number of photons detected. Hence, the resolution improves with increasing fragment length until the resolution is limited by other factors such as flow and laser power instabilities. The opposite effect occurs in electrophoresis where the resolving power decreases with increasing fragment length. Finally, since the fragments are analyzed individually, there is direct accountability of the number of fragments in each identified size class that does not require a tedious calibration.

In the near future it will be possible, in addition to measuring the size of DNA fragments, to determine if a specific sequence is present on the fragment. This will be accomplished by labeled oligonucleotide hybridization. The fluorescent label can be attached externally to the oligonucleotide via a avidin-biotin bond or can be incorporated into the oligo during synthesis. When this measurement is made, flow cytometry will be capable of DNA fingerprinting in a way that is equivalent to electrophoretic techniques. The techniques will be used in a growing number of applications such as DNA fingerprinting for disease diagnostics, forensics, and paternity determination.

DNA fragment size analysis is one form of single molecule detection. In actuality, at a micron per 3 kilobases, the linear dimension DNA fragments can be quite large. When untangled and stretched out a 45 kilobase pair fragment of DNA is 15 microns long. However, unconstrained DNA tends to ball up into an much smaller lump. True single molecule detection has been achieved in a flow cytometer for single fluorescing molecules.

At this point in time several different species of molecules have been detected at the single molecule level. At a molecular weight of 250 kilo Daltons, the largest and the brightest molecule is phycoerythrin (PE) which contains approximately 20 fluorophors. Detection of PE was achieved with flow cytometric techniques similar to those described above for DNA fragment sizing with one added factor. The primary background against which the fluorescence from the PE is detected is Raman scatter of the excitation laser light by water molecules in the sample and the sheath fluids. This interference is eliminated by using time-gated detection of the PE fluorescence excited by a pulsed laser beam. The detection system is gated on after the fast (200 picosecond) light pulse is past to detect the fluorescence photons that are emitted with a lifetime that is typically a few nanoseconds. The intensity of the

Raman scattered light, which has no lifetime, follows the laser excitation pulse in time allowing gating of the detection electronics to record fluorescence photons only during the time that the Raman scattered light is not present. In time-gated fluorescence measurements the detection time relative to the laser light pulse can be measured. By recording the detection time of on the order of 100 photons it is possible to determine the fluorescence decay time of a single molecule(Wilkerson, Goodwin et al. 1993).

Single fluorescent molecules can be distinguished by their emission spectra and by their fluorescence lifetime. Two molecular species have been distinguished by differences in their excitation and emission simultaneously at the single molecule level(Soper, Davis et al. 1992). This capability has led the way to the an ongoing project that seeks to determine the sequence of individual pieces of DNA. The approach is the following. A labeled strand of DNA will be synthesized enzymatically from a template strand with prelabeled nucleotides. Each nucleotide will be labeled with an fluorophor that identifies the nulceotide by the color or the lifetime of the fluorescence emission. A single labeled DNA duplex will be attached to a microsphere and suspended in the flow stream of a flow cytometer capable of single molecule detection and identification. The microsphere will be held in place by either a small pipette or by an optical trap. An exonulcease will be added to the sheath fluid that will release the terminal labeled bases sequentially into the flowing stream. The liberated, labeled bases will be carried through the laser beam(s) where they will be identified in turn and thus the sequence will be determined.

Progress has been made in all phases of this sequencing scheme. Individual molecules have been detected and identified (Ambrose, Goodwin et al. 1993). Single strands of DNA have been attached to microspheres, the microsphere suspended in a flowing fluid, and the DNA observed stretched out in the wake of the bead(Schecker, Goodwin et al. 1995). With multiple fragments of DNA attached to a microsphere, digestion of DNA by an exonuclease in a flowing fluid has been realized by detection of fluorescently labeled bases downstream from the support. Two fluorescently labeled nucleotides have been incorporated enzymatically into bacteriophage M13 DNA out to 7 kilobases. The next step will be to attach one labeled piece of DNA to a microsphere, suspend it in a cytometer, add an exonuclease, and observe photon bursts due to single labeled nucleotides.

There will be two advantages to flow cytometric DNA sequencing, First, it will be rapid. Given the single molecule detection rates that can be achieved and the rate of enzymatic cleavage, it is expected that the sequencing rate will be between 100 and 1000 bases per second. In addition, the length of DNA fragment that can be sequenced in one run will be large, on the order of 40 kilobases. This technique for sequencing DNA will provide a global view of sequence information that is not otherwise available by any other means.

This work is supported by the NIH National Center for Research Resources under Grant RR-01315 for the National Flow Cytometry Resource and by the US Department of Energy, Office of Health and Environmental Research.

Table 1: Range of particles analyzed by flow cytometry

Particle Size (Microns)	Current Capabilities	Future Capabilities
100,000.		Larger plankton
10,000.		
	Zooplankton	
1,000.	Algae	
	Encapsulated cells, bacteria	Bio reactors
100.	Multicellular spheroids	
	Plant protoplasts	Encapsulated cells
		Microcnvironments
10.	Mammalian cells	Cell aggregates
	Nuclei	
	Chromosomes	
1.	Bacteria	
	Picoplankton	
0.1	DNA fragments	
		Smaller DNA fragments
0.01		
	Virus particles	
		Other single molecules
0.001		
	Phycoerythrin	
	Rhodamine	
0.0001		

References:

Alexander, R. G., E. D. Cocking, et al. (1985). "The Characterization and Isolation of Plant Heterokaryons by Flow Cytometry." Protoplasma **128**: 52.

Ambrose, W. P., P. M. Goodwin, et al. (1993). "Application of Single Molecule Detection to DNA Sequencing and Sizing." Ber Bunsenges Phys Chem **97**: 1535.

Chisholm, S. W., R. J. Olson, et al. (1988). "A Novel Free-Living Prochlorophyte Abundant in the Ocean Euphotic Zone." Nature **326**: 655.

Courties, C., A. Vaquer, et al. (1994). "Smallest Eukaryotic Organism." Nature: 255.

Cram, L. S. (1990). "Flow Cytogenetics and Chromosome Sorting." Hum Cell **3**: 99.

Davis, J. H., W. McAllister, et al. (1988). Very Large Particle Sorting Using a Modified FACS-III. XIII International Meeting of the Society for Analytical Cytology, Breckenridge, CO.

Dubelaar, G. B., A. C. Groenewegen, et al. (1989). "Optical Plankton Analyzer: A Flow Cytometer for Plankton Analysis, II: Specifications." Cytometry **10**: 529.

Freyer, J. P., M. E. Wilder, et al. (1987). "Viable Sorting of Intact Multicellular Spheroids by Flow Cytometry." Cytometry **8**: 427.

Glazer, A. N. and H. S. Rye (1992). "Stable Dye-DNA Intercalation Complexes as Reagents for High-Sensitivity Fluorescence Detection." Nature **359**: 859.

Goodwin, P. M., M. E. Johnson, et al. (1993). "Rapid sizing of Individual Fluorescently Stained DNA Fragments by Flow Cytometry." Nucleic Acids Res **21**: 803.

Gorvel, J. P., Z. Mishal, et al. (1988). Characterization of Intestinal Membrane Vesicles with Flow Cytometry. Cell-Free Analysis of Membrane Traffic. New York, NY, Alan R. Liss, Inc.: 195.

Gray, J. W. and L. S. Cram (1990). Flow Karyotyping and Chromosome Sorting. Flow Cytometry and Sorting. M. R. Melamed, T. Lindmo and M. L. Mendelsohn. New York, NY, John Wiley and Sons: 503.

Harkins, K. R. and G. D. W. (1987). "Factors Governing the Flow Cytometric Analysis and Sorting of Large Biological Particles." Cytometry **8**: 60.

Hercher, M., W. Mueller, et al. (1979). "Detection and Discrimination of Individual Viruses by Flow Cytometry." J Histochem Cytochem **27**: 350.

Huller, R., E. Grossner, et al. (1994). "The Macro Flow Planktonmeter: A New Device for Volume and Fluorescence Analysis of Macro Plankton Including Triggered Video Imaging in Flow." Cytometry **17**: 109.

Jett, J. H. and R. G. Alexander (1985). "Droplet Sorting of Large Particles." Cytometry **6**: 484.

Murphy, R. F. (1985). "Analysis and Isolation of Endocytic Vesicles by Flow Cytometry and Sorting: Demonstration of Three Kinetically Distinct Compartments Involved in Fluid-Phase Endocytosis." Proc. Natl. Acad. Sci. USA **82**: 8523.

Nguyen, D. C., R. A. Keller, et al. (1987). "Detection of Single Molecules of Phycoerythrin in Hydrodynamically Focused Flows by Laser-Induced Fluorescence." Analyt Chem **59**: 2158.

Nir, R., R. Lamed, et al. (1990). "Single-Cell Entrapment and Microcolony Development within Uniform Microspheres Amenable to Flow Cytometry." Appl Environ Microbio **56**: 2870.

Peeters, J. C., G. B. Dubelaar, et al. (1989). "Optical Plankton Analyzer: A Flow Cytometer for Plankton Analysis, I: Design Considerations." Cytometry **10**: 522.

Petty, J. T., M. E. Johnson, et al. (1995). "Characterization of DNA Size Determination of Small Fragments by Flow Cytometry." Anal Chem **in press**.

Rens, W., C. H. Van Oven, et al. (1994). "Slit-Scanning Technique Using Standard Cell Sorter Instruments for Analyzing and Sorting Nonacrocentric Human Chromosomes, Including Small Ones." Cytometry **16**: 80.

Sahar, E., R. Nir, et al. (1994). "Flow Cytometric Analysis of Entire Microbial Colonies." Cytometry **15**: 213.

Sanders, C. A., D. M. Yajko, et al. (1991). "Detection and Analysis by Dual-Laser Flow Cytometry of Bacteriophage T4 DNA Inside *Escherichia coli*." Cytometry **12**: 167.

Schecker, J. A., P. M. Goodwin, et al. (1995). Flow-Based Continuous DNA Sequencing Via Single Molecule Detection of Enzymatically Cleaved Fluorescent Nucleotides. Ultrasensitive Instrumentation for DNA Sequencing and Biochemical Diagnostics, San Jose, CA, Society of Photo-Optical Instrumentation Engineers.

Shapiro, H. M. (1994). Practical Flow Cytometry. New York, NY, John Wiley and Sons.

Shera, E. B., N. K. Seitzinger, et al. (1991). "Detection of Single Fluorescent Molecules." Chem Phys Lett **174**: 553.

Soper, S. A., L. M. Davis, et al. (1992). "Identification of Single Molecules in Solution." J Opt Soc Am **B** (: 1761.

Steen, H. B. (1990). Flow Cytometric Studies of Microorganisms. Flow Cytometry and Sorting. M. R. Melamed, T. Lindmo and M. L. Mendelsohn. New York, NY, John Wiley and Sons: 605.

van der Waaij, L. A., G. Mesander, et al. (1994). "Direct Flow Cytometry of Anaerobic Bacteria in Human Feces." Cytometry **16**: 270.

Van Dilla, M. A., R. G. Langlois, et al. (1983). "Bacterial Characterization by Flow Cytometry." Science **220**: 620.

Vesey, G., P. Hutton, et al. (1994). "Application of Flow Cytometric Methods for the Routine Detection of *Cryptosporidium* and *Giardia* in Water." Cytometry **16**: 1.

Weaver, J. C., P. E. Seissler, et al. (1984). "Microbiological Measurements by Immobilization of Cells within Small Volume Elements." Ann N Y Acad Sci **434**: 363.

Weaver, J. C., G. B. Williams, et al. (1988). "Gel Microdroplets: Rapid Detection and Enumeration of Individual Microorganisms by Their Metabolic Activity." Bio/Technology **6**: 1084.

Wilkerson, C. W., P. M. Goodwin, et al. (1993). "Detection and Lifetime Measurement of Single Molecules in Flowing Sample Streams by Laser-Induced Fluorescence." Appl Phys Lett **62**: 2030.

Flow Cytometric Immunophenotyping Using Cluster Analysis And Cluster Editing

Gary C. Salzman, Richard J. Beckman, Jim D. Parson, Aaron M. Nauman, Sigrid J. Stewart, and Carleton C. Stewart
Los Alamos National Laboratory (G.C.S., R.J.B., J.D.P.), Los Alamos, New Mexico, North Carolina State University (A.M.N.), Raleigh, North Carolina, Roswell Park Cancer Institute (S.J.S., C.C.S.), Buffalo, New York

INTRODUCTION

Flow cytometry is the best method for classifying leukemias and lymphomas (Braylan,1989; Braylan, 1993; Deegan, 1989; Drexler, 1987; Duque, 1993; Foon, 1986; Neame, 1986; Stewart, 1989, 1990A, 1990B, Terstappen, 1988; Verwer, 1993; Weber, 1990; Wilman, 1989) . In a flow cytometer, a distinct group of cells is identified by its repertoire of cell surface markers or epitopes. An epitope is labeled using a monoclonal antibody with an attached fluorochrome. A flow cytometer is used to detect the monoclonal antibody through its attached fluorochrome, which, in turn, identifies an epitope on a cell. Appropriate combinations of monoclonal antibodies provide a set of probes for labeling all important epitopes. This process enables the identification of cell types in a specimen. For three-color immunophenotyping, the set of three different probes is referred to as a panel. Each panel is designed to label a specific set of cell types. A panel of three probes provides more information about specific groups of cells than do panels of one or two probes. This additional information requires the use of more complex data analysis methods than are now available.

NATO ASI Series, Vol. H 95
Flow and Image Cytometry
Edited by Alain Jacquemin-Sablon

The positions of the cell clusters in the multidimensional space, as well as, the frequencies of cells within the clusters provide diagnostic information. The position of a cell type-specific cluster is defined by the cluster's centroid. For three-color immunophenotyping, the centroid is defined as a vector with components consisting of the mean values for forward scatter (FSC), side scatter (SSC), and three colors of fluorescence. This cluster centroid position is referred to as a spatial pattern for the cell population, because the centroid is viewed through windows consisting of an array of bivariate dot plots. One sees a set of two dimensional centroid projections for a cluster. The spatial patterns for a heterogeneous population of cells generated by the simultaneous use of three antibodies in normal healthy donors are very consistent among individuals. For example, lymphocytes incubated with fluorochrome-tagged antibodies against CD45, CD14 and HLADr form a tight cluster in the five dimensional space. The frequency of cells in each cluster varies with the donor, but the spatial patterns are constant for healthy donors. This consistency is analogous to the consistency found in histological sections of normal tissue recognized by the pathologist. Large changes in the frequency of cells in normally occurring clusters can be a hallmark of leukemia. Another hallmark is the alteration in these spatial patterns. Leukemic cells often display what is called inappropriate marker expression and lie in abnormal positions in multivariate space. This means that the cells simultaneously express epitopes corresponding to several different cell types. Thus, spatial pattern in addition to the frequency of cells in each cluster becomes diagnostic.

At present, clinical laboratories, even those that do two and three color detection, reduce their data to a single percentage for each antibody when reporting their results. This practice developed from past experience and the strategy built into current data analysis computer programs. Since specific subsets are resolved in normal and leukemic populations, this reduction represents a loss of information that could be useful to the interpretation process of classifying leukemias. Furthermore, normal cells that contaminate leukemic cell populations are often included, leading to misinterpretation of the true leukemic phenotype. Keeping and using more of the multicolor information provides an opportunity to expand the application of flow cytometric immunophenotyping into the management of leukemias through residual disease detection (Ryan, 1993).

Current list mode data analysis programs do not provide adequate tools for separating cell clusters in three-color flow immunophenotyping. A number of these programs have their origin in public domain software written in the late 70s (Salzman, 1978, 1981) when single-color fluorescence was in common use for immunophenotyping. In the current programs, data are presented as multiple bivariate dot plots. The user is able to draw regions around a group of events in a single bivariate dot plot and have events whose components fall inside that region presented in a new color in all of the dot plots. These methods all make cuts orthogonal to the axes with the risk that unseen populations may be included in the counts in the intersecting regions. The Becton-Dickinson Paint-A-Gate program (Terstappen, 1988) tries to deal with this problem by mixing colors in the regions of overlap. These approaches to the analysis of multivariate data are very time consuming and become more difficult as the number of variates increases. Cluster analysis addresses this problem by examining all of the variates simultaneously in isolating cell populations.

Cluster analysis groups similar objects in a multidimensional space. The method examines all of the dimensions in identifying cell clusters. Cluster analysis has been used by a number of investigators to isolate populations of cells in the multivariate flow cytometry data space (Bakker-Schut, 1993; Bartels, 1974; Conrad, 1987; Crowell, 1978; Demers, 1992; Genter, 1979; Kosugi, 1988; Murphy, 1985; Salzman, 1975, 1976, 1990A,1990B, 1993; Terstappen, 1988; Verwer, 1993; Weber, 1990). In immunophenotyping, these approaches are sometimes successful in isolating the cell populations in a panel of antibodies. In our work, a clustering technique called K-means is used in an attempt to isolate specific cell populations. In the case of normal bone marrow, we show that K-means cluster analysis is not completely satisfactory, and we present a cluster editing procedure, which can be used to complete the task.

As three-color flow cytometric immunophenotyping moves into the clinical laboratory, there is an increasing need for rapid and reliable methods to objectively analyze the acquired data. Cluster analysis has been one approach for finding and isolating cell populations in multivariate data . A cluster analysis program has been developed for assisting with the task of partitioning list mode data with an emphasis on three-color human bone marrow immunophenotyping. The program runs on the Macintosh computer. This paper briefly describes the clustering process, and discusses some of the problems with cluster analysis of bone marrow data. Cluster editing is also introduced. This technique enables the user to efficiently move cells from one cluster to another and to combine or separate cells incorrectly assigned by clustering.

MATERIALS AND METHODS

In K-means cluster analysis (Spath, 1985), the user specifies that K clusters are to be located. The algorithm finds K different cluster centroids or means. The goal of the method is to form clusters whose centroids are widely separated and whose variances are small. This section describes the K-means algorithm and two of the distance measures used with it. A simple example using an artificial dataset illustrates the different results obtained using these two distance measures.

The K-means clustering algorithm first assigns each of the N events randomly to one of K clusters. An event is the set of measurements on one cell. The centroid of each of the K clusters is computed. An iterative process begins in which each event, in turn, is moved from cluster to cluster to determine which assignment produces the best fit. Moving an event into a cluster means changing its cluster label from that of one cluster to that of another. The quality of a particular cluster is determined by computing SS, the sum of the squares of the distances between the cluster centroid and the events assigned to the cluster. The overall goodness of this fit is assessed with an objective function, Z, defined as the sum of the SS values for all of the K clusters. For each event, the algorithm finds the cluster for which the value of Z is minimized. The event is appropriately relabeled, and the necessary cluster means are recalculated. This loop through all the events is repeated until fewer than 1% of the events change cluster assignments in an iteration. The mathematical details of this process, along with FORTRAN computer programs, are described in Spath (1985). A graphical illustration of the process is presented in the appendix to Salzman (1993).

Figure 1 illustrates K-means clustering of an artificial data set containing two clusters. The data set was generated using a Gaussian random number generator. The Euclidean distance metric, which is the usual straight line distance measure, is used. Three rows of four bivariate dot plots are shown. FSC is forward scatter, SSC is side scatter, and FL1, FL2 and FL3 are three fluorescence variates. The first row shows all 2400 events. The second row shows cluster 1 with 994 of its actual 1000 events. The third row shows cluster 2 with 1406 events. The actual number of events in cluster 2 is 1400. Figure 1 also shows a departure from the conventional method of presenting clustered data, which is to show all the clusters in different colors overlapped on the same set of bivariate dot plots. This method of display can obscure the visualization of clusters containing very few events. This problem is eliminated when each cluster is shown on its own set of dot plots, so that it can be viewed and analyzed separately. Bakker-Schut (1993) also presents individual clusters on separate plots.

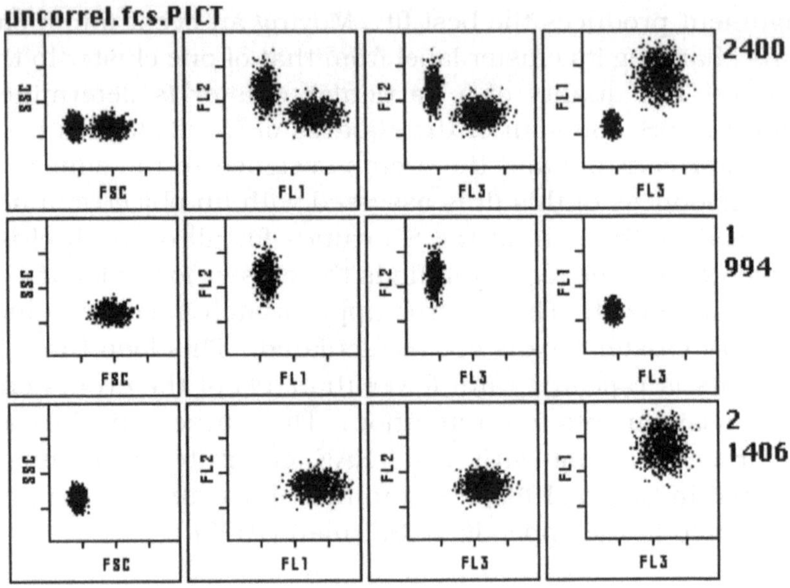

Figure 1. K-means clustering using the Euclidean metric of an artificial data set containing two clusters with 1000 and 1400 events, respectively. Three rows of four bivariate dot plots are shown. FSC is forward scatter, SSC is side scatter, and FL1, FL2 and FL3 are three fluorescence variates. The first row shows all 2400 events. The second row shows cluster 1 with 994 events. The third row shows cluster 2 with 1406 events. Displaying the individual clusters in separate rows facilitates further analysis.

K-means clustering algorithms often use the Euclidean metric to determine the separation between an event and a cluster centroid. While simple to compute, this metric causes the clustering to form hyperspherical clusters with the result that adjacent elliptical clusters are likely to be split in half. Figure 2 illustrates this effect. This example represents a worst case scenario because the two clusters have been created to overlap nearly completely in all the other variates.

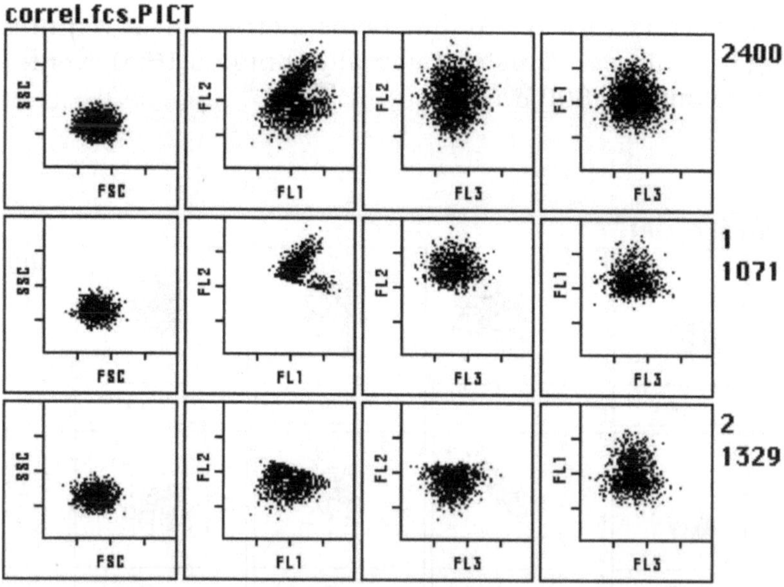

Figure 2. K-means clustering using the Euclidean metric of an artificial dataset containing two clusters with 1600 and 800 events, respectively. Three rows of four bivariate dot plots are shown. In the bivariate display of variates FL1 versus FL2, the clusters appear as adjacent ellipses. The first row shows all 2400 events. The second row shows cluster 1 with 1071 events, and the third row shows cluster 2 with 1329 events. This incorrect clustering illustrates a problem associated with using the Euclidean metric for data sets containing adjacent elliptical clusters. The Euclidean metric tends to force the clustering to form hyperspherical clusters with the result that adjacent elliptical clusters are cut in half.

The problem of cluster splitting is addressed by using the Mahalanobis metric (Mahalanobis, 1936), which takes into account the nonspherical nature of the data in computing distances. This is accomplished by weighting the distance based on the variance-covariance structure of the cluster. Figure 3 shows the same data as in Figure 2 partitioned into two clusters using the Mahalanobis metric. The result of clustering is closer to being correct with the Mahalanobis metric, but it requires 2.5 as long to compute. Cluster 1 actually contains 1600 events, and cluster 2 contains 800 events.

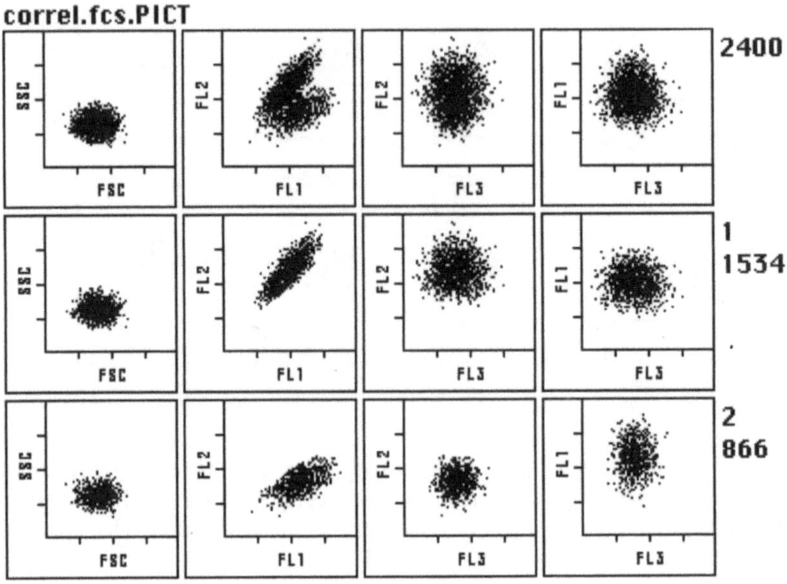

Figure 3. K-means clustering of the data set in Figure 2 using the Mahalanobis metric. The second row shows cluster 1 with 1534 of its 1600 events, and the third row shows cluster 2 with 866 events. The correct number of events is 800. This dataset was specifically designed so that the two clusters differed significantly only in the second bivariate dot plot, which shows the two clusters as adjacent ellipses. The Mahalanobis metric uses the variance-covariance matrix of each cluster to determine the distance of an event from a cluster centroid. This enables the K-means algorithm to more correctly separate the clusters.

In addition to the sensitivity of clustering to geometric shape, K-means clustering fails to find very small clusters. Also, the number of clusters must be specified, or a cluster merging criteria must be established. Thus, K-means clustering alone is inadequate to partition a data set into biologically meaningful populations. This led to the development of the concept of cluster editing, which is illustrated in Figure 4. This process allows events to be moved from one cluster to another by the user. To edit a cluster, the cursor is moved inside one of the bivariate dot plots of the cluster to be edited. A region is then drawn with the mouse around a group of events to be moved out of the cluster. The selected events can then be dragged by the user into another cluster. The points are assigned to the second cluster. This procedure enables close user interaction with the data.

A Macintosh computer program called AutoGate was developed to perform the initial K-means cluster analysis and cluster editing by the user on FCS 1.0 or 2.0 format list mode files. The program enables analysis and editing to be performed in any combination of bivariate views. When cluster analysis is performed, each cluster is assigned a numeric label and a different color. The clusters can be presented on the same display as overlapping groups of colored dots and/or each cluster can be shown on its own set of bivariate displays. Color or gray scale images can be saved as PICT format files and/or printed on Macintosh compatible printers. The statistics for the clusters are presented in a separate window. The statistics included for each cluster are the number of events in the cluster, the means and standard deviations of the variates, and the correlation coefficient matrix. The program is available at no cost from the first author.

The data files for the normal bone marrow specimens discussed here were collected as part of routine clinical laboratory work at Roswell Park Cancer Institute. Preparation and staining of the specimens are described in Stewart (1990A).

RESULTS

How does one determine the correct number of clusters in a complex multivariate dataset? Several authors have addressed

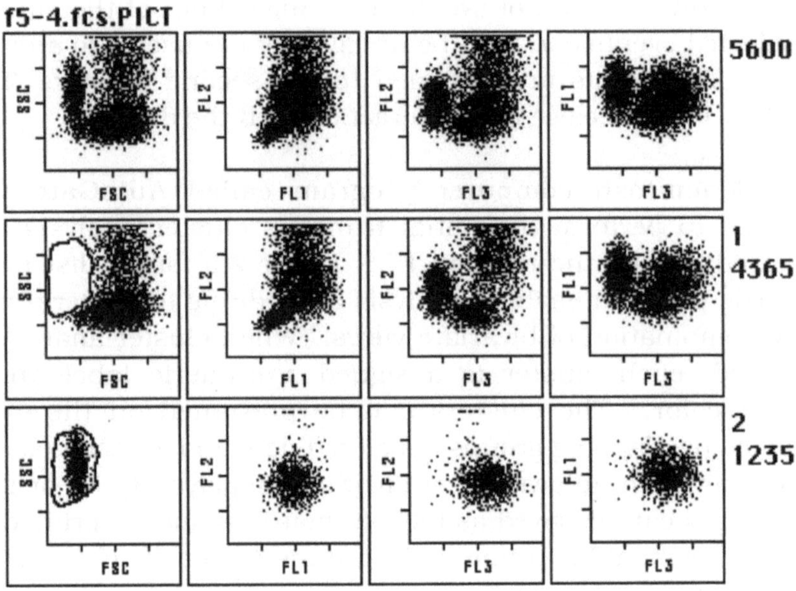

Figure 4. Illustration of cluster editing. An artificial list mode data set containing four populations was clustered with the specification that the clustering algorithm was to find only one cluster but that the display was to provide for two populations. For this case, the clustering algorithm merely copies all the data into cluster one. In the display before editing, cluster one contained 5600 events and cluster two contained zero events. Cluster editing is done in this example by drawing a region around a group of events in cluster one and then dragging those events to cluster two. The figure shows an empty region in cluster one from which the events in cluster two have been taken. Note that the removal of a group of events from cluster one has revealed other clusters that cannot be seen in the display of all the data in the top row. This is particularly evident in column three.

this problem using a variety of statistical tests (Davies, 1979; Milligan, 1985). Most of the tests examine two quantities: an objective function, Z, which is the total within cluster sum of squares, and B, the sum of the squared distances between each cluster centroid and the grand mean of the cluster centroids. The goal of clustering is to form compact clusters that are far apart. This is achieved by minimizing the ratio of Z to B. Davies (1979) computes a quantity that is related to this ratio, here called an R statistic. Figure 5 shows the R statistic as a function of the

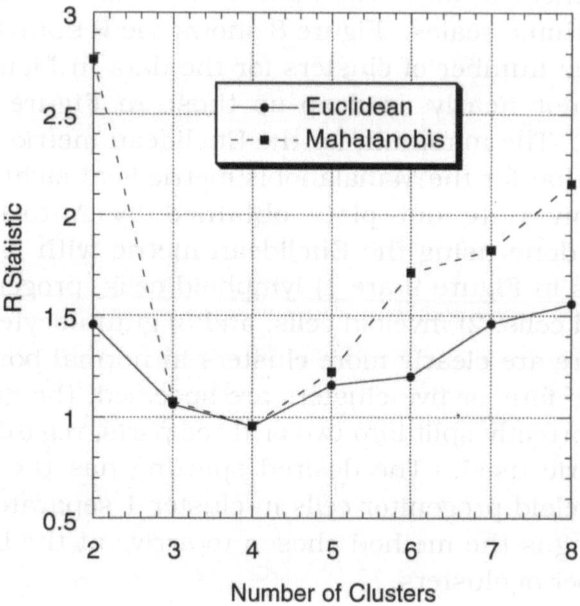

Figure 5. R statistic (Davies, 1979) as a function of the number of clusters for the artificial data set in Figure 4. The solid line is for the Euclidean distance, and the dashed line is for the Mahalanobis distance. Both curves have their minimum at four, which is the correct number of clusters.

assumed number of clusters for the data set in Figure 4. The curve with solid diamonds is for the Mahalanobis metric and that with open squares is for the Euclidean metric. Both curves have a minimum at four clusters, which is the correct number for this data set. Figure 6 shows the result of K-means clustering on this data set using the Mahalanobis metric with K = 4.

Analyzing data from three-color immunophenotyping of bone marrow specimens represents a significant departure from the comfortable world of normally distributed artificial data sets. Figure 7 shows four bivariate dot plots from one three-color immunophenotyping panel for normal human bone marrow. This panel is intended to enable the determination of a leukocyte differential. The forward scatter (FSC) and side scatter (SSC) variates are linear scales. The fluorescence variates are all four decade logarithmic scales. Figure 8 shows the R statistic (9) as a function of the number of clusters for the data in Figure 7. The minima are not nearly as deep as those in Figure 6 for the artificial data. The minimum for the Euclidean metric is at three clusters and that for the Mahalanobis metric is at eight. Figure 9 shows the bivariate dot plots obtained by AutoGate when clustering is done using the Euclidean metric with K = 3. The three clusters in Figure 9 are 1) lymphoid cells, progenitor cells, and red blood cells, 2) myeloid cells, and 3) granulocyte-like cells. However, there are clearly more clusters in normal bone marrow than three. If four or five clusters are specified, the granulocyte cluster is incorrectly split into two or three parts, regardless of the distance metric used. The desired splitting has the T cells, B cells, and myeloid progenitor cells in cluster 1 separated instead. Cluster editing is the method chosen to arrive at the biologically correct number of clusters.

The strategy for each three-color panel is to use cluster analysis to partition the data into a "reasonable" number of clusters and then edit those clusters to arrive at a final partition of the data set. Figure 9 shows a normal bone marrow data set partitioned into three clusters using the Euclidean metric. Two empty rows are provided for cluster editing. This initial clustering separates the groups of cells sufficiently that simple cluster editing can be used to quickly complete the task. The results of editing are shown in Figure 10. The Mahalanobis metric with K = 8 was also used for clustering the data in Figure 9. This

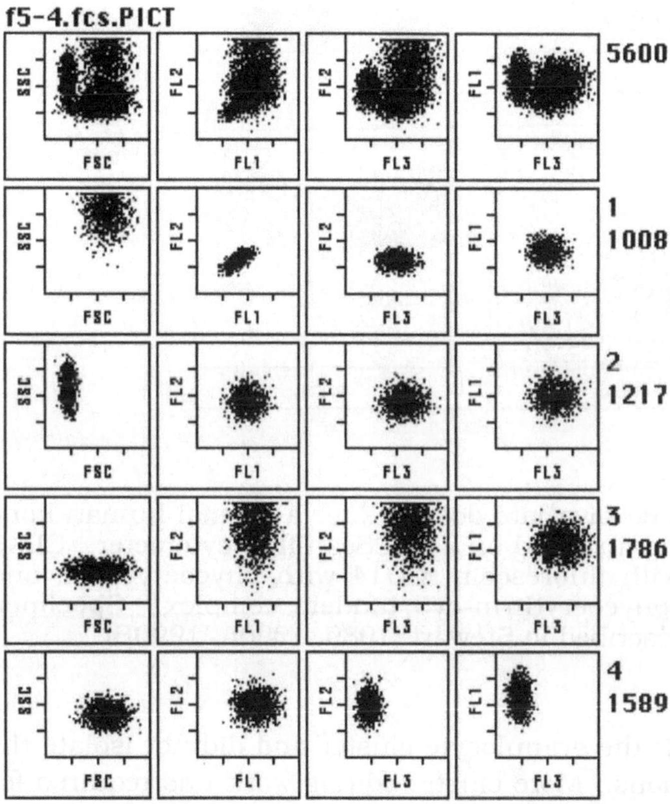

Figure 6. K-means cluster analysis with K = 4 of an artificial data set containing four clusters. The Mahalanobis metric was used. The correct number of events in each of clusters one through four are 1000, 1200, 1800, and 1600, respectively. The numbers of events found by AutoGate in these clusters are 1008, 1217, 1786 and 1589, respectively.

NormalBoneMarrow.PICT

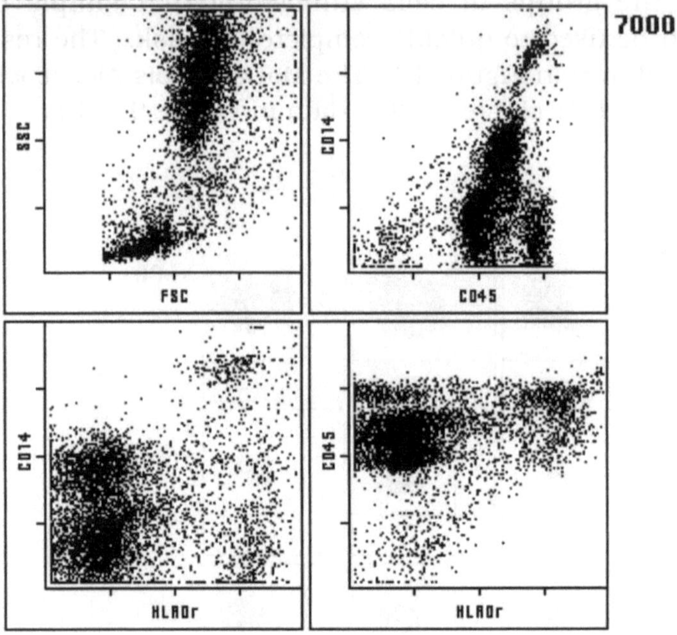

Figure 7. Selected bivariate dot plots for a normal human bone marrow specimen analyzed on a FACScan flow cytometer. CD45 is conjugated with fluorescein, CD14 with phycoerythrin, and HLADr with a phycoerythrin-cy5 tandem complex. Specimen preparation is described in Stewart (1989, 1990A, 1990B).

partitioning split the granulocyte cluster and did not isolate the desired populations. More cluster editing work was required for this approach than for the case in which the Euclidean metric with K = 3 was used.

The clustering of the normal bone marrow data in Figure 7 provides an initial objective partitioning of the multivariate data. This makes the subsequent cluster editing much easier and more efficient. One could bypass the clustering step and carry out the entire partitioning process using only cluster editing. However,

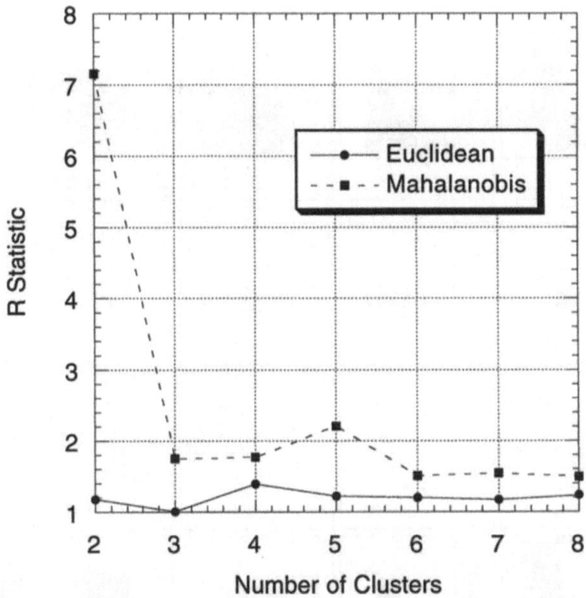

Figure 8. R statistic (Davies, 1979) as a function of the number of clusters for the normal human bone marrow data shown in Figure 7. The solid line is for the Euclidean distance, and the dashed line is for the Mahalanobis distance. The minimum R for the Euclidean metric is at three clusters, while that for the Mahalanobis metric is at eight clusters. These are very shallow minima, which indicate that the cluster means are close together and that the cluster variances are large.

the order in which the editing is done then becomes important as does the skill of the individual doing the editing. The cluster / edit method provides significant improvement over current methods regardless of the skill of the user.

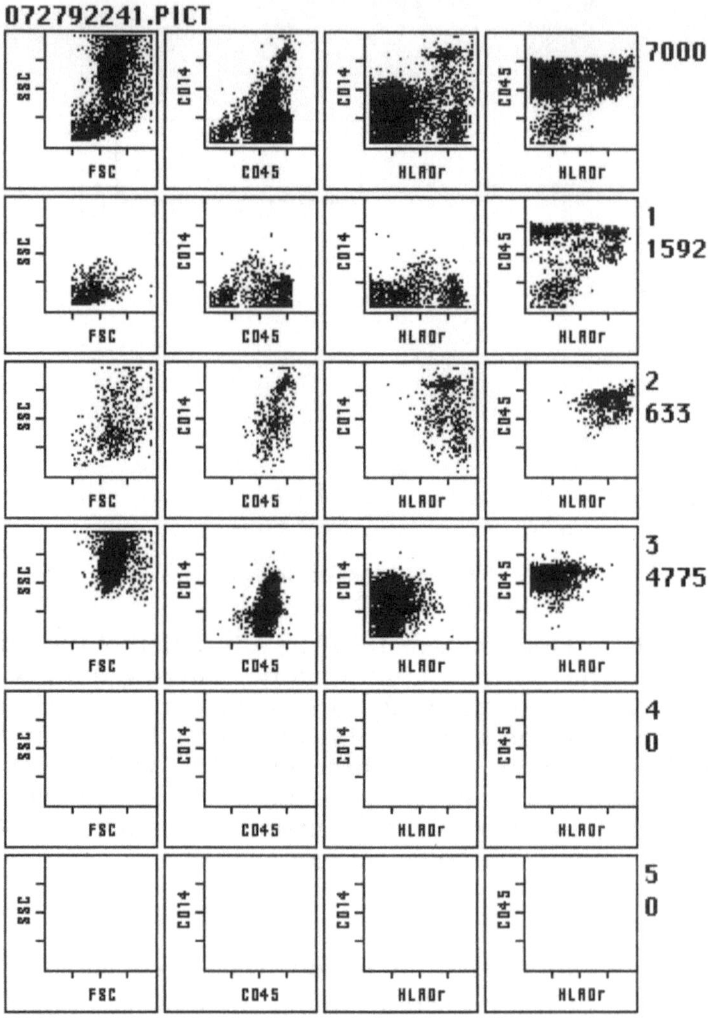

Figure 9. Normal bone marrow data set clustered using the Euclidean metric with K = 3. The top row shows the three clusters overlaid. Rows numbered 1, 2 and 3 contain the three clusters. Cluster 1 contains lymphoid cells and red blood cells, cluster 2 contains myeloid cells, and cluster 3 contains granulocyte-like cells. Two additional empty clusters are shown to provide space in which to do cluster editing.

DISCUSSION

K-means cluster analysis is useful for an initial objective partitioning of multivariate flow cytometry data. K-means cluster analysis alone, however, is insufficient to partition a human bone marrow data set into a biologically correct set of classes. Cluster editing enables a user to efficiently complete the partitioning task once an initial clustering is done. Cluster analysis combined with cluster editing enables the correct partitioning of normal bone marrow data sets from three-color immunophenotyping.

Now what does one do with a data set from a patient with some type of leukemia? Such a data set contains some cells corresponding to those present in a normal bone marrow data set. The data set also contains abnormal cells that have no such correspondence. One would like to determine the immunophenotype of the abnormal cells in the patient bone marrow specimen. A method is needed to define the boundaries of each cluster in the normal bone marrow, so that unknown cells located outside the boundaries of clusters containing normal cells can be classified. The technique of Classification And Regression Trees (CART) (Breiman, 1984) is an appropriate method for classifying patient bone marrow data sets and is discussed in Beckman (1995).

Ten panels of three antibodies each are used for flow immunophenotyping of bone marrow specimens at Roswell Park. For each panel, a normal bone marrow template is developed by clustering and editing and then combining several data sets. Each normal bone marrow panel template is used to train a CART classifier for that panel. The training produces a set of hyper planes that effectively enclose each cluster in the normal bone marrow template. The clusters are separated from each other and

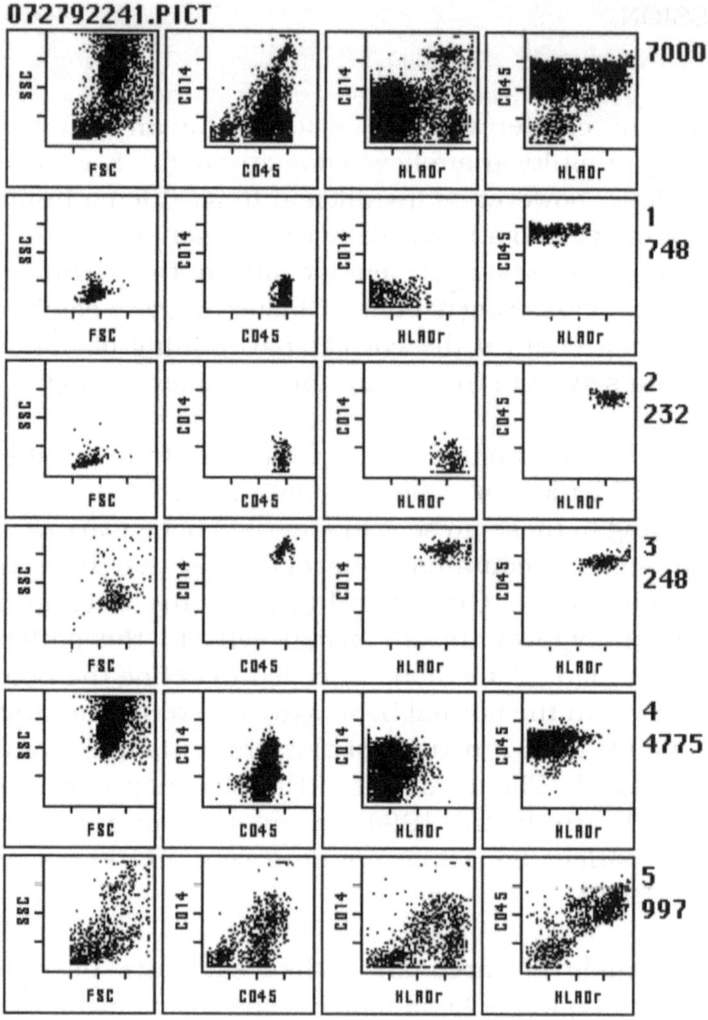

Fig. 10. Result of cluster editing the normal bone marrow data in Figure 9. The cluster contents are: 1) T and NK cells, 2) B cells, 3) monocytes, 4) granulocytes, and 5) other, which includes red cells and progenitor cells not specifically identified by this three-color panel.

from the surrounding empty space. The empty space is defined as a cluster containing cells that exhibit inappropriate antibody expression. The set of hyperplanes for each panel is used to classify bone marrow data sets from leukemia patients. Some of the patient bone marrow cells fall into the inappropriate expression cluster. The immunophenotype of this cluster can be determined easily.

ACKNOWLEDGMENTS

This work was performed under the auspices of the U.S. Department of Energy at the Los Alamos National Laboratory and under the auspices of the New York State Department of Health at Roswell Park Cancer Institute and was supported by U.S. National Institutes of Health. grants CA54518, RR01315, and the New York State Department of Health. The authors wish to thank Ann Nettleton for her assistance in acquiring the flow cytometry data.

REFERENCES

Bakker Schut TC, De Grooth BG, Greve J. Cluster analysis of flow cytometric list mode data on a personal computer. Cytometry 14:649-659, 1993.

Bartels PH, Olson GB, Jeter, WS, Wied GL: Evaluation of unsupervised learning algorithms in computer analysis of lymphocytes. Acta Cytol. 18:376-388, 1974.

Beckman RJ, Salzman GC, Stewart CC: Classification and Regression Trees for Bone Marrow Immunophenotyping. Cytometry, 1995, in press.

Braylan RC, Benson NA: Flow cytometric analysis of lymphoma. Arch. Pathol. Lab. Med. 113:627-633, 1989.

Braylan RC: Lymphomas. In: Clinical Flow Cytometry. Principles and Applications, Bauer KD, Duque RE , Shankey TV (eds). Williams and Wilkins, Baltimore, 1993, pp. 203-234.

Breiman L, Friedman JH, Olshen RA, Stone CJ: Classification and Regression Trees. Wadsworth and Brooks, Monterey, CA , 1984.

Conrad MP: A rapid, nonparametric clustering scheme for flow cytometry data. Pattern Recog. 20:229-235, 1987.

Crowell JM, Hiebert RD, Salzman GC, Price BJ, Cram LS, Mullaney PF: A light-scattering system for high-speed cell analysis. IEEE Trans. Biomed. Engin. BME-25:519-526, 1978.

Davies DL, Bouldin DW: A cluster separation measure. IEEE Transactions on Pattern Analysis and Machine Intelligence. PAMI-1:224-227, 1979.

Deegan MJ: Membrane antigen analysis in the diagnosis of lymphoid leukemias and lymphomas. Arch. Pathol. Lab. Med. 113:606-618, 1989.

Demers S, Kim J, Legendre P, Legendre L: Analyzing multivariate flow cytometric data in the aquatic sciences. Cytometry 13: 291-298, 1992.

Drexler HG: Classification of acute myeloid leukemias - a comparison of FAB and immunophenotyping. Leukemia 1: 697-705, 1987.

Duque RE: Acute Leukemias. In: Clinical Flow Cytometry. Principles and Applications, Bauer KD, Duque RE , Shankey TV (eds). Williams and Wilkins, Baltimore, 1993, pp. 235-246.

Foon KA, Todd RF: Immunologic classification of leukemia and lymphoma. Blood 68: 1-31, 1986.

Genter FC, Salzman GC: A statistical approach to the classification of biological cells from their diffraction patterns. J. Histochem. Cytochem. 27:268-272, 1979.

Kosugi Y, Sato R, Genka S, Shitara N, Takakura K: An interactive multivariate analysis of FCM data. Cytometry 9:405-408, 1988.

Mahalanobis PC: On the generalized distance in statistics. Proc. Natl. Inst. Sci. India 2:49-55, 1936.

Milligan GW, Cooper MC: An examination of procedures for determining the number of clusters in a data set. Psychometrika 50: 159-179, 1985.

Murphy RF: Automated identification of subpopulations in flow cytometric list mode data using cluster analysis. Cytometry 6:302-309, 1985.

Neame PB, Soamboonsrup P, Browman GP, Meyer RM, Benger A, Wilson WE, Walker IR, Saeed N, McBride JA: Classifying acute leukemia by immunophenotyping: a combined FAB - immunologic classification of AML. Blood 68: 1355-1362, 1986.

Ryan DH: Detection of minimal residual disease by flow cytometry. In: Clinical Flow Cytometry. Principles and Applications, Bauer KD, Duque RE , Shankey TV (eds). Williams and Wilkins, Baltimore, 1993, pp. 479-496.

Salzman GC, Crowell JM, Goad CA, Hansen KM, Hiebert RD, LaBauve PM, Martin JC, Ingram M, Mullaney PF: A flow-sytem multiangle light scattering instrument for cell characterization. Clin. Chem. 21:1297-1304, 1975.

Salzman GC, Crowell JM, Hansen KM, Ingram M, Mullaney PF: Gynecologic specimen analysis by multiangle light scattering in a flow system. J. Histochem. Cytochem. 24:308-314, 1976.

Salzman GC, Hiebert RD, Crowell JC: Data acquisition and display for a high speed cell sorter. Comput. Biomed. Res. 11: 77-88, 1978.

Salzman GC, Wilkins SF, Whitfil JA. Modular computer programs for flow cytometry and sorting. Cytometry 1: 325-336, 1981.

Salzman GC, Stewart CC, Duque RE: Expert systems for flow cytometry data analysis: a preliminary report. In: New technologies in cytometry and molecular biology. Salzman GC (ed). Proc. SPIE 1206:98-105, 1990A.

Salzman GC, Duque RE, Braylan RC, Stewart CC: A CLIPS expert system for clinical flow cytometry data analysis. In: Proceedings of the 1st CLIPS user's group conference, Houston, Texas, 13-15 August 1990B.

Salzman GC, Bartels PH: Expert systems for cytometry data analysis. In: Clinical Flow Cytometry. Principles and Applications, Bauer KD, Duque RE , Shankey TV (eds). Williams and Wilkins, Baltimore, 1993, pp. 585-601.

Spath H: Cluster dissection and analysis. Theory, FORTRAN programs and examples. Halsted Press. John Wiley and Sons, New York, 1985.

Stewart CC: Multiparameter flow cytometry for clinical applications. In: New technologies in cytometry. Salzman GC (ed). Proc. SPIE 1063:150-160, 1989.

Stewart CC: Cell preparation for the identification of leukocytes. In: Methods in Cell Biology, Volume 33, Flow Cytometry. Darzynkiewicz Z, Crissman C (eds). Ch. 39. Academic Press, San Diego, 1990A.

Stewart CC: Multiparameter analysis of leukocytes by flow cytometry. In: Methods in Cell Biology, Volume 33, Flow Cytometry. Darzynkiewicz Z, Crissman C (eds). Ch. 40. Academic Press, San Diego, 1990B.

Terstappen LWMM, Loken MR: Five-dimensional flow cytometry as a new approach for blood and bone marrow differentials. Cytometry 9:548-556, 1988.

Verwer BJH, Terstappen WMM: Automatic lineage assignment of acute leukemias by flow cytometry. Cytometry Suppl. 6: 44, 1993.

Weber JE, Bartels PH: Statistical identification of subpopulations for flow cytometric data. In: New technologies in cytometry and molecular biology. Salzman GC (ed). Proc. SPIE 1206:19-30, 1990.

Wilman C, Stewart CC: General principles of multiparameter flow cytometric analysis: applications of flow cytometry in the diagnostic pathology laboratory. Semin. Diagn. Pathol. 6:3-12, 1989.

Introduction to High-Speed Flow Sorting

J.F. Keij
New York Blood Center
310 East 67th Street
New York, NY 10021

General Introduction

Purification of viable cells by flow cytometric cell sorting, initiated in the mid sixties, has become standard procedure in a variety of research fields. Single cell suspensions, which are stained with fluorochromes to distinguish desired from undesired cells, can be sorted into two purified populations. Sorts can be performed under sterile conditions and usually the viability of the sorted cells is excellent.

Cell sorters have been used successfully in research fields ranging from microbiology and botany to andrology, hematology, and immunology. Cells have been sorted for strain improvement (Betz *et al.*, 1990), characterization of plant protoplasts (Galbraith *et al.*, 1988), isolation of X- or Y-bearing sperm cells (Johnson *et al.*, 1989), microscopic observation (Alberti *et al.*, 1984) and functional characterization of hematological cell types (Shah *et al.*, 1986). For most of these applications a few hours of sorting suffices, as relatively small samples are required.

However, some applications require longer sort times. The isolation of the rare pluripotent hematopoietic stem cells for in vivo transplantation in mice (Visser *et al.*, 1984) which requires the processing of large numbers of cells, takes all day. Extrapolating this technique to human applications would require sort times in excess of several days. Although alternative techniques involving anti-CD34 beads allow rapid isolation of hematopoietic cells, the purity of these cell fractions may not be sufficient (Berenson *et al.*, 1991). Whereas one of the new high-speed sorters could purify the pre-enriched CD34 cell fraction within 2 hours, current droplet sorters would require 22 hours to perform this task.

NATO ASI Series, Vol. H 95
Flow and Image Cytometry
Edited by Alain Jacquemin-Sablon

As stated above, purging of undesired cells from bone marrow samples intended for transplantation requires the processing of large numbers of cells. Bulk irradiation of photosensitized bone marrow samples containing acute myeloid leukemia (AML) cells has been shown to reduce the colony forming ability of the AML progenitor cells by more than 90 % (Singer *et al.*, 1988). However, the colony forming ability of the normal myeloid progenitor cells (GM-CFC) was also reduced by 40 %. High-speed droplet or photodamage cell sorting of AML samples could increase the selective purging ratio, and thus provide better grafts for transplantation.

Another application that requires high sort rates is the sorting of X- or Y-bearing sperm for in vitro insemination. Other techniques, such as density gradients (Upreti *et al.*, 1988), do not yield fractions enriched for either X- or Y-bearing sperm. The insemination of one cow requires 5 x 10^6 sorted sperm cells (Morrell *et al.*, 1988) and only minor populations should be sorted to obtain highly purified fractions of X- or Y-bearing sperm cells (Johnson *et al.*, 1989). Again, a high-speed sorter could sort sufficient numbers of X- or Y-bearing sperm to make insemination of pre-sexed sperm a realistic application. Furthermore, the reduced sort times would significantly improve the viability of the sorted fractions.

The sorting of rare hybridoma class-switch variants, that occur at frequencies of approximately 10^{-5} to 10^{-6} (Dangl *et al.*, 1982), is another application that requires the processing of large numbers of cells to obtain sufficient variants for several selection rounds.

It can be concluded that the applications mentioned above become feasible when a fast flow sorter or sorting technique is applied. Before deciding on the type of cell sorter or sorting technique to use, the specific sort rate limiting features of the different available flow sorters should be analyzed. Three critical stages in the sort process can be distinguished. The first stage is the *measurement* of the particle that enters into the laser beam. The *processing* of the detected signal is the second stage, and the execution of the *sort* decision represents the final and most important rate limiting step.

Detecting a particle

In a flow cytometer, cells flow through a focused light beam in single file. Light scatter and fluorescence from each individual cell are collected by detectors. For accurate light measurements, the sample must remain optimally aligned with respect to the illumination source and the light collecting detectors. For this purpose the sample is injected into a fast flowing sheath fluid. Laminar flow of the sample fluid and sheath fluid in the emerging fluid jet fixes the sample position to the center of the fluid jet (Figure 1).

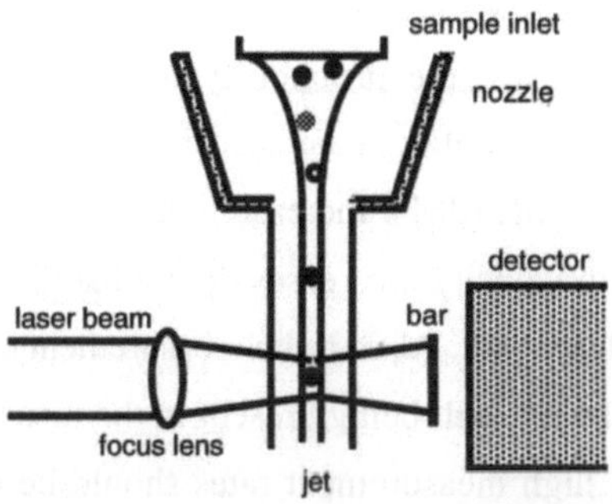

FIG. 1. A schematic representation of the detection of cells in a flow cytometer. Cells, hydrodynamically focused to the center of the fluid jet, pass through a focused laser beam. Light signals are collected by detectors. The signal to noise ratio is improved by blocking the laser beam from reaching the detectors.

Sheath fluid, usually a buffered saline solution, is filtered before use to eliminate small light scattering debris. The optically clean sheath fluid is delivered to a cuvette or a nozzle tip from a storage container. Sheath fluid velocities of 2 and 10 m/s are routine for cytometers equipped with a cuvette or a nozzle tip, respectively. Samples are either injected into the sheath fluid by a mechanically driven syringe or by positive air pressure.

Argon-ion lasers are commonly used to excite the fluorochromes that identify the desired cell population. These lasers offer a stable output, sufficient power and are easy to focus onto the sample using a single lens.

The polarized laser beam is focused onto the sample area forming a bright round spot of 20 - 50 μm in diameter. Due to the gaussian intensity profile of the laser light spot, slight deviations in the position of the sample can result in significant deviations in the measured fluorescence intensities. Applications that require a high degree of accuracy, such as DNA-ploidy analysis, have led to the use of cross cylindrical lenses. These lenses yield wider, elliptical spots, so the generated signal is less sensitive to the position of the cells within the horizontal plane of the sample core.

As a cell enters the laser spot it scatters laser light in all directions. The forward scatter detector registers photons scattered along the propagation direction of the laser beam. Light scattered under a 90° angle is collected by the side scatter detector. Optical filters are inserted in front of the respective detectors to spectrally separate the emissions from different fluorochromes.

As the cell rate increases, the probability of detecting a cell in the laser spot increases. However, the probability of a detected event consisting of more than one cell also increases (Lindmo *et al*, 1981). Increasing the average distance between the cells, by increasing the sheath velocity, decreases the probability of detecting multiple cells simultaneously in the laser spot. Also, if the measurement area is decreased in size, the probability of more than one cell being present in the measurement area, i.e., the laser spot, decreases. Thus, high measurement rates should be achievable in a flow sorter with a small laser spot height and a high sheath velocity.

These options are currently offered in two commercial flow cytometers from Becton Dickinson (San Jose, CA) and Cytomation (Fort Collins, CO). However, during the development of a high-speed photodamage cell sorter, it was found that not all cells survive passage through a nozzle at increased sheath fluid velocities ranging from 20 - 36 ms/s (Keij *et al*., 1994). The damage, predominantly to larger cells, was caused by shear forces that the cells endure during acceleration in the nozzle tip. Currently, there are no published data on the survival of large cells that were sorted with these commercial high-speed sorters, using increased sheath fluid velocities.

Three additional, instrument related parameters influence the level of coincident arrival in the laser spot. In most commercial instruments the signal-to-noise ratio is improved by filtering the noise component. Unfortunately, this results in wider signal pulses and thus in a greater probability of pulse overlap (coincidence). To determine the amount of pulse shaping, one can simply compare the measured pulse width with calculated laser transit time (= {cell diameter + spot height} / sheath fluid velocity). Secondly, pulse integration also leads to longer pulses and should be avoided. Finally, attempts to eliminate detected doublets from the sample requires the use of pulse width detectors. In general the application of such detectors results in a small reduction of the overall processing rate as the instrument processing time is increased when these detectors are activated. As the purity of the desired population normally is of prime importance, this small processing rate trade-off should be accepted.

Processing the detected particle

The intense forward scatter (FSC) signal is focused onto a photodiode, whereas the much less intense side scatter and fluorescence signals are focused on photomultipliers. The currents generated by photons hitting the detectors are converted to voltage signals and can be amplified.

To start the processing of the signals from the detectors, the electronics need to be triggered. Normally a FSC threshold is used, because all cells passing through the laser spot generate a strong FSC signal. When the FSC signal exceeds the preset threshold value, the electronics are triggered and the peak-sense-and-hold modules start to follow the incoming signals. After passage of the cell through the laser beam the FSC signal drops below the threshold value and the peak values are retained. The peak values of all signals are then processed by the analog-to-digital converters (ADC). The digitized signals are finally transferred to a storage medium for analysis or are used to specify the criteria for sorting.

High sort rates can be achieved when the time needed to process a particle is minimal. This time interval is referred to as the processing dead time, which normally ranges from 6 to 50 µs, depending on the instrument. In high-speed instruments, this dead time is kept to a minimum by applying high-speed electronics, and reducing the amount of intentional signal pulse shaping.

Flow sorters equipped with more than one laser for detection, normally have a processing dead time greater than single laser cytometers. After processing the signals from the first laser the sort decision is delayed until the data from the second laser are processed. In the mean time the electronics usually cannot process incoming particles. Thus, the highest processing rates are achieved with sorters equipped with high-speed electronics and only one laser for detection.

Parallel processing of the incoming data (Van den Engh and Stokdijk, 1989) is a method to achieve high processing rates for multi-laser cytometers. The processing dead time in such a system is independent of the number of parameters and detection lasers. The ADC, which represents the slowest component (4 - 6 µs) in such a system, determines the maximum processing rate (Keij *et al.*, 1991). This concept has been applied in the MoFlo, a high-speed droplet sorter, and in a high-speed photodamage cell sorter (Keij et al., 1995). Processing rates in excess of 40,000 events/s are common.

Sorting the processed particle

After the processed data are compared to the sort criteria, the sort decision is made: the cell should or should not be sorted. However, the execution of the sort decision must be delayed until the cell reaches the sort area of the flow system. As the cell is about to be sorted some instruments will check whether the desired cell should actually be sorted or not. This depends on the level of purity that is desired. If highly purified sorted populations are demanded then it is essential that the sheath fluid volume containing the sorted cell does not contain other cells. In case an unwanted cell was detected close enough to the sorted cell to be included in the sort volume the sort electronics will abort execution of the sort decision. As the time to sort a desired cell normally represents the longest time interval in the process of sorting, it is this interval that is of critical importance in high-speed sorting: the deflection time must be kept to a minimum.

Different flow cytometric principles have been used to purify cells. In the following paragraphs several sorter types and their rate limitations will be briefly discussed.

fluid switch sorters

Although droplet sorters are the most common, fluid switch sorters are used increasingly to isolate small samples and to sort biohazardous samples. In one such cuvette based sorter, the desired cells were separated from the sample stream by triggering a vacuum switch which allowed the desired cells to be isolated by suction (Kamenski and Melamed, 1967). Sort rates of up to 300 cells/s could be achieved. Acoustic transducers (Friedman, 1974; Sweet, 1979; Dühnen *et al*., 1983) have also been used to isolate the desired cells at sort rates of up to 500 cells/s. Although relatively slow, fluid switch sorters are the instruments of choice for the sorting of infectious material as this can be done safely in a completely enclosed environment. Commercial fluid switch sorters that are capable of isolating 300 cells/s, such as the

FACScan and the two-laser Excalibur (Becton Dickinson) are available for exactly that purpose. In these instruments, cells are collected/sorted by moving an aspiration needle from the sheath fluid into the sample core to collect the desired cell. As the needle continuously collects fluid, sorted samples are extremely diluted.

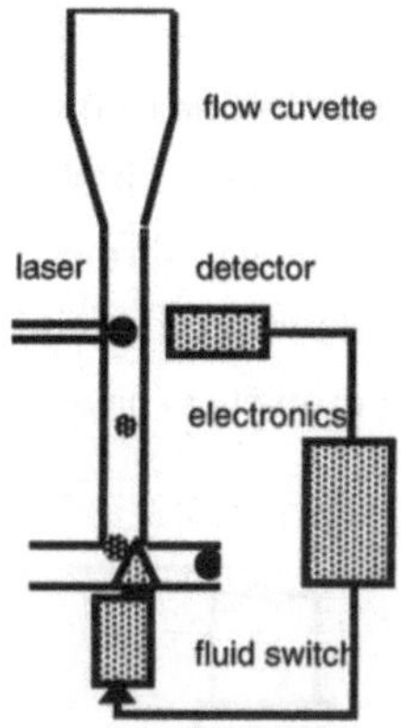

FIG. 2. A diagram of a fluid switch sorter. Cells are detected in the laser beam. The sort decision that is formulated in the electronics is executed by displacing the fluid switch at the end of the flow cuvette. Desired cells are deflected to the left, and all other cells are deflected to the right.

Besides providing a safe and simple sorting platform, sorting of large particles (0.1 - 1 mm) can only be done with fluid switch sorters. A fluid switch option for the cuvette based Optical Plankton Analyzer (Figure 2) for the sorting of very long and fragile algae (Dubelaar *et al.*, 1989; Peeters *et al.*, 1989) has recently been completed. In the Optical Plankton Analyzer the cuvette ends in a T-piece that contains a wedge that can be manipulated by a piezo crystal. Desired cells can be deflected in one direction, while undesired cells flow in the other direction.

A major drawback of fluid switch sorters is the fact that they can only sort one population at a time. Furthermore, due to the mechanical constraints of the sorting process fluid switch sorters will never be able to sort at rates higher than several hundred cells per second. However, when the frequency of the cells of interest is low, these sorters allow for 'high-speed' isolation of such rare cells. For example, stem cells in cord blood represent 0.1 - 2.0 % of a Ficoll separated fraction. These cells can be effectively sorted at throughput rates of 10,000 - 15,000 cells/s, if the system is triggered by the fluorescence of the anti-CD34 antibody that is used to identify the stem cells (McCoy *et al.* 1991). Although this method is fast and requires no operator supervision, it will require a second 'clean-up' sort at a low detection rate to eliminate any contaminating CD34- cells that were not detected during the fast sort.

droplet sorters

Most cell sorters employ droplet deflection (Fulwyler, 1965; Sweet, 1979) to sort desired cells. In a droplet sorter, the fluid jet is broken into droplets by vibrating the nozzle tip assembly at a constant frequency. Two deflection plates, charged to around +1500 and -1500 volts are situated downstream. Initially, the entire jet is maintained at zero volts, allowing the uncharged droplets to pass down between the charged plates undeflected (Figure 3).

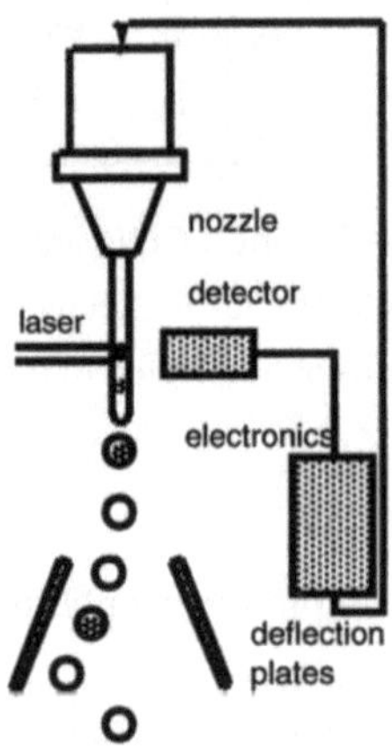

FIG. 3. A diagram of a droplet sorter. The fluid jet emerging from the nozzle is broken up into droplets. The sort decision that is formulated in the electronics is executed by charging the droplets containing the desired cells. The droplets containing the desired cells are deflected as they pass the deflection plates. Droplet sorters can sort two populations: one to the left and the other to the right. Undesired cells pass between the plates without deflection and are collected in a waste container.

Two sort windows can be set up for the simultaneous sorting of two populations. After each cell is analyzed, the sorter electronics check whether the cell belongs to one of these two populations. If so, a sort pulse is programmed to charge the fluid jet at a preset delay time. This delay time is equal to the time it takes for the cell to travel from the laser-jet intersection to the droplet break-off point of the fluid jet. By charging the fluid jet, and thus the droplet breaking off from the jet to either +100 or -100 volts, the droplet will be electrically charged and deflected towards the oppositely charged plate. Unsorted cells are discarded in a waste container. The sorted cells are generally collected into tubes, which can be cooled for improved viability. Current droplet sorters can reliably sort cells at rates of up to 4,000 cells/s and achieve purities exceeding 97 %.

In droplet sorters, the sort rate is limited by the number of droplets generated per second. A droplet sorter operating at a droplet frequency of 40,000/s, which represents a sort dead time of 25 μs, could theoretically deflect 40,000 cells/s. Slight deviations in the sample flow, due to partial obstructions in the nozzle orifice, may result in a desired cell not arriving at the break-off point in time. This will lead to the deflection of an empty droplet and a reduction of the sort yield. Increasing the number of droplets in the deflection envelope to two or three droplets per sorted cell increases the probability of deflecting the selected cell. However, the sort dead time will also increase, to 50 and 75 μs, respectively, and the maximum sort rates will consequently be reduced to 20,000 and 13,300/s.

If the purity of the sorted cells has to be high, then the maximum sort rates must be much lower to reduce contaminating passenger cells in the deflected droplets. To obtain sorted fractions of a purity greater than 90 %, the maximum measurement rates were calculated for a FACS II sorter (Visser and Van den Engh, 1982) to be 4,200, 1,400 and 800 cells/s for droplet envelopes containing 1, 2 and 3 droplets, respectively.

Circuits that abort the sort decision when coincidence is detected in the measurement, the processing and/or the sort time intervals, are helpful to increase the level of purity of the sorted population. However, inclusion of such aborts in the sorting decision results in lower yields. Fine tuning the balance between yield and purity can be done experimentally or mathematically by using models that incorporate the detection dead time, the processing dead time, the sort dead time, the detection rate and the sort rate (Keij *et al.*, 1991; Leary *et al.*, 1994).

Although droplet sorters can sort much faster than fluid switch sorters, sort rates exceeding 10,000 cells/s are not realistic. To overcome the sort rate limitations imposed by the droplet frequency, one can opt to sort without coincidence checking or opt for the previously mentioned fluorescence trigger method. In both cases the yield is optimal, but a second round of slow sorting is required to improve the purity of the sorted population.

high-speed droplet sorters

The Lawrence Livermore National Laboratory High-Speed Sorter (Peters *et al.*, 1985) was built for the high-speed isolation of chromosomes. The droplet frequency was increased from 40,000/s to 200,000/s, which reduced the deflection dead time from 25 μs to 5 μs. To obtain these high droplet frequencies the sheath fluid velocity was increased to 50 m/s. Sort rates in excess of 20,000 cells/s were reported, but cell survival was poor due to the high sheath fluid velocity.

Improved energy transfer between the droplet generating piezoelectric crystal and the fluid jet led to a sorter that could be operated a lower sheath fluid pressures (Farmer *et al.*, 1991). Droplet frequencies of 150,000/s were achieved at normal sheath fluid pressures, and measurement and sort rates of 15,000/s were reported. Data on the sorting of viable cells is not available.

The Rochester High-Speed Sorter (Leary *et al.*, 1991; Leary *et al.*, 1994) was designed for rare event analysis and enrichment sorting. The instrument operates under more or less standard conditions: a moderate sheath fluid velocity of 20 m/s is used and sorting is performed at a droplet frequency of 30,000 droplets/s. Using single droplet deflection, sorts at measurement rates in excess of 100,000 cells/s have been reported. As enriching the rare cells for further analysis was the main goal, the expected contaminations were considered acceptable (Leary *et al.*, 1994).

Recently, two commercial high-speed sorters have become available. The MoFlo which is based on the parallel processing design (Van den Engh and Stokdijk, 1989) that was described earlier, employs moderate sheath fluid pressures and sorts at measurement rates of up to 20,000 cells/s. A similar instrument was developed at Systemix (Palo Alto, CA) for the high-speed isolation of human bone marrow stem cells for transplantation purposes. Both instruments operate at moderate sheath fluid pressures and the survival of sorted lymphocytes is good. No survival data on larger cells is currently available.

The second commercial high-speed droplet sorter consists of a FACS Vantage (Becton Dickinson) that is modified to operate at increased droplet frequencies and increased sheath fluid velocities. At this time no published data on sort yield, purity and cell survival are available.

electroporation sorters

The sorters in the previous paragraphs sort desired cells by physically removing them from the sample: zappers are a different breed. These instruments do not remove desired cells from the sample but *kill the undesired* cells. The first zapper that successfully sorted cells in flow (Bakker Schut *et al.*, 1990) electrocuted undesired cells as they flowed through a Coulter orifice. The reversing pulses with field strengths of up to 3 x 10^6 V/m were applied for a duration of 200 μs to ensure complete destruction of the undesired cells. Sort purities of 95 % at cell rates of up to 5,000/s were achieved. Unless the duration time of the zap-pulse can be considerably reduced, electroporation sort rates will be similar to the sort rates of commercial droplet sorters.

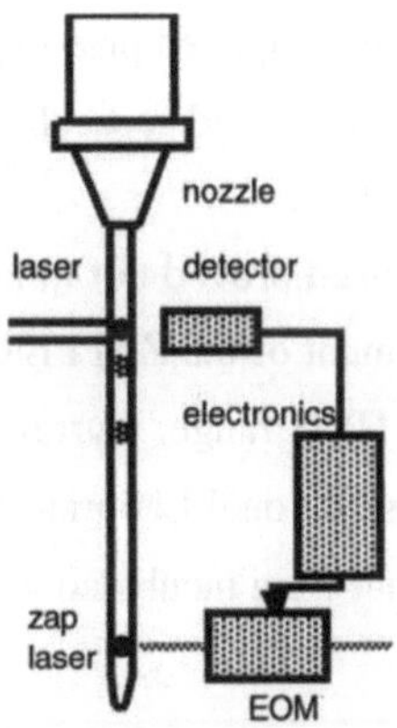

FIG. 4. A diagram of a photodamage sorter. The sort decision is executed by deflecting the zap laser beam onto the undesired cells, while desired cells are allowed to pass the zap laser station unharmed. All cells are collected in the same vessel.

photodamage sorters

A photodamage sorter (Martin and Jett, 1981; Shapiro, 1983; Herweijer, 1988) is equipped with two lasers. One laser is for the detection of the cells and the other is for the lethal irradiation of the undesired cells (Figure 4). The principle was first tested using a modified droplet sorter, the ZAPPER (Herweijer *et al.*, 1988), and the execution of the sort decision was be performed by an Acousto-Optic Modulator (AOM) which could deflect the zap laser beam within 0.2 μs. Thus, the sort time interval was reduced to 0.4 μs. The ZAPPER represented the first sorter that could sort cells faster than it could measure them. Thus, sort rates were now limited by the dead time of the particle measurement and signal processing.

In the first experiments, cells were cultured with 5-bromo-2'-deoxyuridine (BrdUrd) for 48 hours and then incubated with Hoechst 33342. Out of 100,000 cells only one cell survived (Herweijer *et al.*, 1988) exposure to 300 mW laser light (353/364 nm) that was focused to a 50 μm spot in the flow cytometer. Although limited to the sorting of actively dividing cell samples, this work presented an important breakthrough for high-speed sorting.

To overcome the limitations imposed by the use of Hoechst 33342/BrdUrd, the beam of a frequency doubled argon ion laser was used to kill the undesired cell, and deflection of the 257 nm laser beam was accomplished with an Electro Optic Modulator (EOM). Testing of the completed prototype (Keij *et al.*, 1995) revealed that the system, with a dead time of 4 μs, could sort cells at rates of 20,000 - 40,000 cells/s. Yields and purities were around 90%.

However, the EOM that was used proved not to be the optimal device to switch the UV laser beam and the development of the ZAPPER will be continued using an AOM with good transmission in the UVC range. Currently, a similar instrument is under development in the Los Alamos National Laboratory. This instrument will be used to photoinactivate chromosomes that were incubated with psoralen.

Summary

Relatively high-speed isolation of (rare) cells can be achieved with commercial instruments if the critical time intervals in the sorting process are kept to a minimum. For routine droplet sorting this translates to sorting cells using a small laser spot, a sheath fluid velocity as high as possible (50 μm nozzle tip), a single laser, no pulse processing and no coincidence aborts. To further increase the throughput rate, the critical fluorescence parameter can be used to trigger the system.

For increased purity, pulse processing and coincidence abort circuits should be activated and the consequent loss in cell yield must be accepted. In case the fluorescence trigger method is used, a slow re-sort will yield highly purified populations.

Fluid switch sorters are able to handle rare event sorting at fairly high processing rates, but a second round of slow sorting is required to purify the sorted population. Droplet sorters are more suited for the isolation of larger populations as they are generally capable of executing the sort decision at 10-fold higher frequencies. Photodamage cell sorters may prove to be the ultimate speed demons but they are still under development.

Before investing in a commercial high-speed cell sorter of any type, it is crucial to verify that the viability of the cells of interest is acceptable. For example, hybridoma cells, liver cells, sperm cells and plant protoplasts will not survive passage through a nozzle at high sheath fluid velocities. Although the initial interest may be lymphocytes exclusively, experience has taught that a good flow cytometry facility will attract users from all 'walks of life'.

Literature cited

Alberti S, Stovel R, Herzenberg LA: Preservation of cells sorted individually onto microscope slides with a fluorescence-activated cell sorter. Cytometry 5:644-647, 1984.

Bakker Schut TC, de Grooth BG, Greve J: A new principle of cell sorting by using selective electroporation in a modified flow cytometer. Cytometry 11:659-66, 1990.

Berenson RJ, Bensinger WI, Hill RS, Andrews RG, Garcia-Lopez J, Kalamansz DF, Still BJ, Spitzer G, Buckner CD, Bernstein ID, Thomas ED: Engraftment after infusion of CD34+ marrow cells in patients with breast cancer of neuroblastoma. Blood 77:1717-1722, 1991.

Betz JW, Aretz W, Härtel W: Use of flow cytometry in industrial microbiology for strain improvement programs. Cytometry 5:145-150, 1984.

Dangl JL, Parks DR, Oi VT, Herzenberg LA: Rapid isolation of cloned isotype switch Variants variants using fluorescence activated cell sorting. Cytometry 2:395-401, 1982.

Dubelaar GJB, Groenewegen AC, Stokdijk W, Van den Engh GJ, Visser JWM: Optical Plankton Analyzer: a flow cytometer for plankton analysis, II: specifications. Cytometry 10: 529-539, 1989.

Dühnen J, Stegeman J, Wiezorek C and Mertens H: A new fluid switching flow cytometer. Histochemistry 77:117-121, 1983.

Farmer C, Esposito R, Van den Engh G: High speed chromosome sorting with a “jet in air" nozzle. Cytometry Supplement 5:116, 1991.

Friedman M: Digital fluidic amplifier particle sorter. U.S. Patent 3,791,517, 1974.

Fulwyler MJ: Electronic separation of biological cells by volume. Science 150:910-911, 1965.

Galbraith DW, Harkins KR, Jefferson RA: Flow cytometric characterization of the chlorophyll contents and size distributions of plant protoplasts. Cytometry 9 :75-83, 1988.

Herweijer H, Stokdijk W, Visser JWM: High-speed photodamage cell selection using bromodeoxyuridine/Hoechst 33342 photosensitized cell killing. Cytometry 9:143-149, 1988.

Herweijer H: Photodamage cell selection: Development of a high-speed cell selection method based on flow cytometry. Thesis, University of Leiden, Leiden, The Netherlands, 1988.

Johnson LA, Flook JP, Hawk HW: Sex preselection in rabbits: live births from X and Y sperm separated by DNA and cell sorting. Biol Reprod 41:199-203, 1989.

Kamenski LA and Melamed MR: Spectrometric cell sorter. Science 156:1364-1365, 1967.

Keij JF, Groenewegen AC, Stokdijk W, Visser JWM: Coincidence in high-speed flow cytometry: models and measurements. Cytometry 12:398-404, 1991.

Keij JF, Groenewegen AC, Visser JWM: High-speed photodamage cell sorting: an evaluation of the ZAPPER prototype. In: Methods in Cell Biology, VOL. 42. Darzynkiewicz Z, Robinson JP, Crissman HA (eds.). Academic Press, San Diego, 1994, pp 371-386.

Keij JF, Groenewegen AC, Dubelaar GBJ, Visser JWM:High-speed photodamage cell selection using a frequency doubled argon ion laser. Cytometry 19:209-216, 1995.

Leary JF, Ellis SP, McLaughlin SR, Corio MA, Hespelt S, Gram JG, Burde S: High-resolution separation of rare cell types. In: Cell Separation Science and Technology (ACS symposium series 464). Kompala DS and Todd P (eds). ACS, Washington, 1991, pp 26-40.

Leary JF: Strategies for rare cell detection and isolation. In: Methods in Cell Biology, VOL. 42. Darzynkiewicz Z, Robinson JP, Crissman HA (eds.). Academic Press, San Diego, 1994, pp 331-358.

Lindmo T, Fundingsrud K: Measurement of the distribution between cell passages in flow cytometry as a method for the evaluation of sample preparation procedures. Cytometry 2, 151-154, 1981.

Martin JC, Jett JH: Photodamage, a basis for super high speed cell selection.

Cytometry 2:114, 1981.

Morrell JM, Keeler KD, Noakes DE, Mackenzie NM, Dresser DW: Sexing of sperm By flow cytometry. Veterinary Record 122:322-324, 1988.

McCoy JP, Chambers WH, Lakomy R, Campbell JA, Stewart CC: Sorting minor subpopulations of cells: use of fluorescence as the triggering signal. Cytometry 12:268-274, 1991.

Peeters JCH, Dubelaar GJB, Ringelberg J, Visser JWM: Optical Plankton Analyzer: a flow cytometer for plankton analysis, I: design considerations. Cytometry 10:522-528, 1989.

Peters D, Branscomb E, Dean P, Merrill T, Pinkel D, Van Dilla M, Gray J: The LLNL high-speed sorter: design features, operational characteristics, and biological utility. Cytometry 6:290-301, 1985.

Shah PD, Keij J, Gilbertson SM, Rowley DA: Thy-1+ and Thy-1- natural killer cells. J Exp Med. 163:1012-1017, 1986.

Shapiro HM: Apparatus and method for killing unwanted cells. U.S. Patent 4,395,397, 1983.

Singer CRJ, Linch DC, Bown SG, Huehns ER, Goldstone AH: Differential phtalocyanine photosensitization of acute myeloid leukaemia progenitor cells: a potential purging technique for autologous bone marrow transplantation. Brit J Haematol 68:417-422, 1988.

Sweet RG: Flow sorter for biological cells. In: Flow Cytometry and Sorting. Melamed MR, Mullaney PF and Mendelson ML (eds). Wiley, New York, 1979, pp. 177-189.

Upreti GC, Riches PC, Johnson LA: Attempted sexing of bovine spermatozoa by fractionation on a percoll density gradient. Gemete Research 20:83-92, 1988.

Van den Engh G, Stokdijk W: Parallel processing data acquisition system for Multilaser multilaser flow cytometry and cell sorting. Cytometry 10: 282-293, 1989.

Visser JWM, Bauman JGJ, Mulder AH, Eliason JF, Leeuw AM: Isolation of murine pluripotent hemopoietic stem cells. J. Exp. Med. 59:1576-1590, 1984.

Visser JWM, Van den Engh GJ: Immunofluorescence measurements by flow cytometry. In: Immunofluorescence Technology, Wick G, Traill KN, and Schauenstein K (eds). Elsevier Biomedical Press, Amsterdam, The Netherlands, 1982, pp. 95-128.

Applications of Near-field Microscopes to Cell Biology

Jacques Barbet, Jean Thimonier and José Rocca-Serra
CNRS, GDR 976
Parc Scientifique et Technologique de Luminy, Case 901
13288 Marseille CEDEX 9, FRANCE

Introduction

Near-field microscopes are based on various physical principles but they all involve scanning a macroscopic tip over the observed sample. In fact, an important part of the discovery made by Binnig and Rohrer (Nobel Prize in physics in 1986 for the scanning tunneling microscope, Binnig and Rohrer, 1982) was that piezoelectric devices are capable of controlled movements in the sub-nanometer range. Today, in near-field microscopy, a single piezoelectric tube is responsible for tip (or sample) scanning and for the control of the distance between the tip and the sample.

There are indeed a variety of near-field microscopes. Essentially any local physical phenomenon may be used to develop a near-field microscope. The key is the term "near-field". It means that the monitored tip-sample interaction is not wave-propagated. As a result, Rayleigh diffraction does not apply and resolution is theoretically unlimited. Examples are tunneling current (scanning tunneling microscope: STM, Binnig and Rohrer, 1982), a quantum phenomenon occurring when two conducting solids are close enough to mix their electronic orbitals, contact interaction (atomic force microscope: AFM, also called scanning force microscope: SFM, Binnig et al., 1986), where the relationship between force and distance is described by the well-known Morse curve, and light in the near-field, also called evanescent, mode (Pohl, 1986). They all have attracted interest in a wide variety of domains and have been applied to biological samples. Optical near-field microscopes (photon scanning tunneling microscope: PSTM and scanning near-field optical microscope: SNOM) have not yet received as much applications in biology. They will have important applications though, given the possibility of monitoring local absorption spectra and local fluorescence, possibly down to the single molecule. This introduction to near-field microscopy in cell biology will however deal

NATO ASI Series, Vol. H 95
Flow and Image Cytometry
Edited by Alain Jacquemin-Sablon

mostly with the atomic force microscope (AFM) which is of utmost interest in the study of cells.

The STM and the AFM

To detect a tunneling current, a voltage bias, usually from a few millivolts to about one volt, is applied between a sharp metal tip and the sample. The tip is then approached toward the sample, first mechanically then by the piezoelectric tube, up to a point where a measurable current (usually from a few picoamperes to several nanoamperes) is detected. The tip-sample distance at this point is in the nanometer range. The tunneling current varies exponentially with the tip-sample distance and it is then easy to keep this distance constant by means of an electronic (analog or digital) feed-back loop while the tip is scanned over the sample. One then obtains the topography of electron density which is the STM image of the sample. The instrument is capable of atomic resolution (Hamers et al., 1986). The limitation is that electric conductivity is required. In chemistry, small molecules are imaged with great resolution, especially when ordered arrays of material adsorbed on the substrate (usually graphite) are formed (Smith et al., 1989). For instance, individual methylene groups of alkanes are imaged and averaging is not necessary. For larger and less ordered systems, the technique becomes more difficult. In biology progresses have been relatively slow but images of macromolecules of biological interest have been published providing useful structural information. Developments are made in two directions. First, STM operation at low currents (a few picoamperes) allowed us and others to observe individual molecules. For instance, the variability in three-dimensional structure and flexibility of molecules such as DNA (Marilley et al., 1994, Pasero et al., 1994) or proteins of the immunoglobulin super-family (Rocca-Serra, et al., 1994, Thimonier et al., 1994) was studied. Chemically modified substrates (especially gold) which provide stronger substrate adhesion (Bottomley et al., 1993) improve the reproducibility of the observation of biological molecules with the STM. The other route is the use of higher voltage bias (a few volts) and extremely low current (fractions of picoamperes) which apparently makes it possible to use the STM on insulating samples and substrate, the conductivity being provided by adsorbed water molecules (Guckenberger et al., 1994). Thus the STM is useful, and will become even more so, for studying molecules, from chemistry to biology. However imaging intact, living cells, does not seem possible at present with this instrument.

The AFM does not require electric conduction. A small microfabricated tip, attached to a cantilever, is used to gently press on the surface of the sample. The cantilever has a force constant in the range of 0.1 N/m. To provide resonance frequencies higher than 10 KHz their mass is reduced to less than a nanogram. The contact force bends the cantilever of which bending and torsion are monitored by reflecting a low-energy laser beam (generated by a small laser-diode) on the cantilever to a four-quadrant light detector. Again, by keeping the difference signal constant during scanning, one gets an image of sample topography. By combining various scanning directions and the possibility to monitor independently the bending and the torsion of the cantilever, images in various modes (topography, friction, error signal) may be recorded.

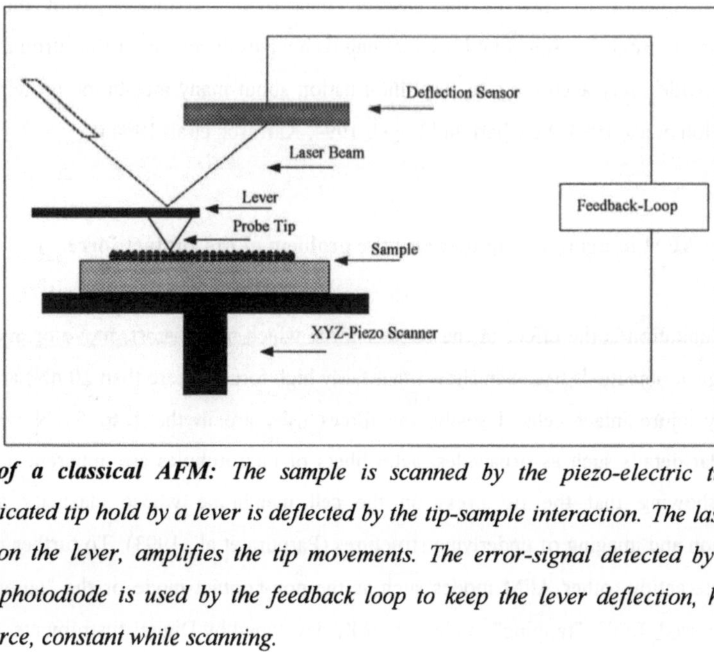

Schema of a classical AFM: *The sample is scanned by the piezo-electric tube. The microfabricated tip hold by a lever is deflected by the tip-sample interaction. The laser beam, reflected on the lever, amplifies the tip movements. The error-signal detected by the four quadrant photodiode is used by the feedback loop to keep the lever deflection, hence the applied force, constant while scanning.*

Imaging biological molecules with the AFM

The absence of a requirement for conductivity has attracted a lot of interest on this technique for biological applications. Biological molecules in water adsorb on mica or glass making possible the observation with the AFM. Again, membranes (Hansma et al., 1991), isolated proteins (Ill et al., 1993, Karrasch et al., 1994b, Thimonier et al., in press), DNA (Hansma et al., 1993, Bezanilla et al., 1993), DNA-protein complexes (Rees et al., 1993) have been studied. Resolution is usually not as good as that of the STM but reproducibility is much greater. An example of the kind of information provided by the AFM is the study of DNA bending by RNA polymerases (Rees et al., 1993). Among the most remarkable achievements of the AFM in this field is the observation of two-dimensional crystals of proteins embedded in phospholipids. The stability of the crystal arrangements provide a very good resolution, sometimes superior to that of the electron microscope, associated with direct three-dimensional information. Applying the averaging techniques developed for electron microscopy images provides very accurate structural information about many membrane proteins (see for instance Hoh et al., 1993, Schabert and Engel, 1994, Karrasch et al, 1994a).

Improved AFM imaging techniques and the problem of the contact force

A major limitation is the effect of the contact force which may distort, move or even destroy the biological objects. It has been shown that fairly high forces (more than 10 nN) are required to grossly injure intact cells. Usually, the forces used are in the 1 to 5 nN range. Then intracellular details such as organelles, actin fibers or microtubules are apparent in the AFM images, showing that the tip press on the cell membrane whose elasticity allows for deformation and imaging of underlying structures (Parpura et al., 1993). To further reduce tip-sample interaction, other AFM modes such as the non-contact mode or the "tapping" mode have been used. In the "tapping" mode, originally developed by Digital Instruments, the tip (or the sample) is vibrated at high frequency and large amplitude. When the tip-sample distance is reduced the tip begins to interact with the sample and the amplitude of the vibration is reduced. The variation of this amplitude is then used by the feed-back loop. The resulting tip-sample interaction is not precisely known but the applied force seems to be reduced and in particular the drag force is annihilated. Under these conditions the cell cytoskeleton is not apparent anymore and the AFM allows for an examination of the unperturbed cell membrane (Putman et al., 1994).

Measurement of receptor-ligand interaction forces

The interaction force is not solely a limitation for the AFM. Monitoring the friction may be used to distinguish between various substrate coating. In addition, the AFM may also be used in biophysics to monitor interaction forces between receptors and ligands. Here, there is no imaging but the minimum force required to detach, for instance, biotin from avidin has been precisely measured at the level of one individual complex (Florin et al., 1994). Several groups are working with tips covered with various ligands to monitor specific interactions on cell surfaces.

Observation of living cells

Another very exciting avenue opened by the AFM is the observation of live cells. The AFM may be operated under liquids and especially under physiological buffers. At present, this still requires a complicated experimental set-up to control the temperature and other parameters, but the results are fascinating. Live cells are observed with a resolution close to that of the scanning electron microscope and cell kinetics may be monitored on individual cells imaged from time to time, an image being recorded in a few seconds. No special substrates are required: glass or tissue culture plastic are commonly used. This opens up a new area for cell biology where very discrete events will be viewed directly without the difficult recreation of a sequence of events from pictures of different cells. The technique is still in a process of validation but it will become broadly available very soon. Today, one of the most spectacular example has been the movie of a virus budding out of a cell shown in real time (Häberle et al., 1992). Many other reports have shown high resolution images of fixed cells (Braunstein and Spudich, 1994), live cells (Henderson et al., 1992) and dynamic studies of cellular activity (Beckman et al., 1994, Fritz et al., 1994, Schoenenberger and Hoh, 1994, Bonfiglio et al., 1995). Preliminary reports show that immunogold staining is possible with the AFM as it is with the electron microscope (Putman et al., 1993, Eppel et al., 1995). The AFM tip may be used to locally injure the observed cells (Parpura et al., 1993) but the effect of scanning on the metabolism of these cells remains to be evaluated.

Conclusion

As a conclusion, I would say that near-field microscopy still requires a great deal of dedication to obtain good images. On-going developments will make the instrument more user-friendly and techniques to prepare and observe the samples will become mature. The experience of many years, which has been acquired in electron microscopy for instance, is still missing. But progresses are made and the usefulness of the new microscopes is now demonstrated. Applications to molecular biology as well as to the study of cells is rapidly becoming very important. The coupling of the AFM with optical microscopes, especially the confocal microscope will make it a very useful instrument. Other technologies such as the near-field optical microscopes will also play an important role in image cytometry. All these techniques are very complementary and they will be used simultaneously to solve new questions in cell biology.

References

Beckmann M, Kold HA and Lang F (1994) Atomic force microscopy of peritoneal macrophages after particle phagocytosis. J Membrane Biol 140:197-204

Bezanilla M, Bustamante CJ and Hansma HG (1993) Improved visualization of DNA in aqueous buffer with the atomic force microscope. Scanning Microsc 7:1145-1148

Binnig G and Rohrer H (1982) The scanning tunneling microscope. Helv Phys Acta 55:726-733

Binnig G, Quate CF and Gerber Ch (1986) Atomic force microscope. Phys Rev Lett 56:930-933

Bonfiglio A, Parodi MT and Tonini GP (1995) Subcellular details of early events of differentiation induced by retinoic acid in human neuroblastoma cells detected by atomic force microscope. Exp Cell Res 216:73-79

Bottomley LA, Haseltine JN, Allison DP, Warmack RJ, Thundat T, Sachleben RA, Brown GM, Woychik RP, Bruce-Jacobson K and Ferrel TL (1993) Scanning microscopy of DNA: the chemical modification of gold for immobilization of DNA. J Vac Sci Technol A10:591-595

Braunstein D and Spudich A (1994) Structure and activation dynamics of RBL-2H3 cells observed with scanning force microscopy. Biophys J 66:1717-1725

Eppell SJ, Simmons SR, Albrecht RM and Marchant RE (1995) Cell-surface receptors and proteins on platelet membranes imaged by scanning force microscopy using immunogold contrast enhancement. Biophys J 68:671-680

Florin EL, Moy VT and Gaub HE (1994) Adhesion forces between individual ligand-receptor pairs. Science 264:415-417

Fritz M, Radmacher T and Gaub HE (1994) Granula motion and membrane spreading during activation of human platelets imaged by atomic force microscopy. Biophys J 67:36-41

Guckenberger R, Heim M, Ceuc G, Knapp HF, Wiegräbe W and Hillebrand A (1994) Scanning tunneling microscopy of insulators and biological specimen based on lateral conductivity of ultrathin water films. Science 266:1538-1540

Häberle W, Hörber JKH, Ohnesorge F, Smith DPE and Binnig G (1992) In situ investigations of single living cells infected by viruses. Ultramicroscopy 42-44:1161-1167

Hamers RJ, Tromp RM and Demuth JE (1986) Surface structure of Si(111)-(7x7) resolved in real space. Phys Rev Lett 56:1972-1975

Hansma HG, Bezanilla M, Zenhausern F, Adrian M and Sinsheimer RL (1993) Atomic force microscopy of DNA in aqueous solutions. Nucl Acid Res 21:505-512

Hansma HG, Weisenhorn AL, Edmunson AB, Gaub HE and Hansma PK (1991) Atomic force microscopy: seeing molecules of lipid and immunoglobulin. Clin Chem 37/9:1497-1501

Henderson E, Haydon PG and Sakaguchi DS (1992) Actin filament dynamics in living glial cells imaged by atomic force microscopy. Science 257:1944-1946

Hoh JH, Sosinsky GE, Revel JP and Hansma PK (1993) Structure of the extracellular surface of the Gap junction by atomic force microscopy. Biophys J 65:149-163

Ill CR, Keivens VM, Hale JE, Nakamura KK, Jue RA, Cheng S, Melcher ED, Drake B and Smith MC (1993) A COOH-terminal peptide confers regiospecific orientation and facilitates atomic force microscopy of an IgG1. Biophys J 64:919-924

Karrasch S, Dolder M, Schabert F, Ramsden J and Engel A (1993) Covalent binding of biological samples to solid supports for scanning probe microscopy in buffer solution. Biophys J 65:2437-2446

Karrasch S, Hegerl R, Hoh JH, Baumeister W and Engel A (1994a) Atomic force microscopy produces faithful high-resolution images of protein surfaces in an aqueous environment. Proc Natl Acad Sci USA 91:836-838

Karrasch S, Heins S, Aebi U and Engel A (1994b) Exploring intermediate filaments structure with the scanning force microscope: comparison with transmission electron microscopy data. J Vac Sci Technol B12:1474-1477

Marilley M, Pasero P, Humbert A, Granjeaud S, Dayez M, Pierrisnard R and Jordan B (1994) Scanning tunneling microscopy study of a DNA fragment of known size and sequence. Microsc Microanal Microstruct 5:47-56

Parpura V, Haydon PG and Henderson E (1993) Three-dimensional imaging of living neuron and glia with the atomic force microscope. J Cell Sci 104:427-432

Pasero P, Blettry C, Marilley M, Jordan B, Granjeaud S, Dayez M and Humbert A (1994) Scanning tunneling investigation of DNA structures involved in gene regulation. J Vac Sci Technol B12:1521-1525

Pohl DW (1986) Near-field scanning optical microscopy. IBM J Res Dev 30:417-419

Putman AJ, de Grooth BG, Hansma PK, van Hulst NF and Greve J (1993) Immunogold labels: cell surface markers in atomic force microscopy. Ultramicroscopy 48:177-182

Rees WA, Keller RW, Vesenka JP, Yang G and Bustamante C (1993) Evidence of DNA bending in transcription complexes imaged by scanning force microscopy. Science 260:646-1649

Rocca-Serra J, Thimonier J, Chauvin JP and Barbet J (1994) Scanning tunneling microscopy of proteins of the immunoglobulin super-family. J Vac Sci Technol B12:1490-1493

Schabert FA and Engel A (1994) Reproducible acquisition of Escherichia coli porin surface topographs by atomic force microscopy. Biophys J 67:2394-2403

Schoenenberger CA and Hoh JH (1994) Slow cellular dynamics in MDCK and R5 cells monitored by time-lapse atomic force microscopy. Biophys J 67:929-936

Smith DPE, Hörber H, Gerber C and Binnig G (1989) Smectic liquid crystal monolayers on graphite observed by scanning tunneling microscopy. Science 245:43-45

Thimonier J, Chauvin JP, Barbet J and Rocca-Serra JP (1994) Scanning tunneling microscopy of monoclonal immunoglobulin G. Microsc Microanal Microstruct 5:341-349

Thimonier J, Chauvin JP, Barbet J and Rocca-Serra J. Preliminary studies of an immunoglobulin M by near-field microscopies. J Trace Microprobe Tech, in press

SUBJECT INDEX

Springer-Verlag and the Environment

We at Springer-Verlag firmly believe that an international science publisher has a special obligation to the environment, and our corporate policies consistently reflect this conviction.

We also expect our business partners – paper mills, printers, packaging manufacturers, etc. – to commit themselves to using environmentally friendly materials and production processes.

The paper in this book is made from low- or no-chlorine pulp and is acid free, in conformance with international standards for paper permanency.

NATO ASI Series H

Vol. 1: **Biology and Molecular Biology of Plant-Pathogen Interactions.** Edited by J.A. Bailey. 415 pages. 1986.

Vol. 2: **Glial-Neuronal Communication in Development and Regeneration.** Edited by H.H. Althaus and W. Seifert. 865 pages. 1987.

Vol. 3: **Nicotinic Acetylcholine Receptor: Structure and Function.** Edited by A. Maelicke. 489 pages. 1986.

Vol. 4: **Recognition in Microbe-Plant Symbiotic and Pathogenic Interactions.** Edited by B. Lugtenberg. 449 pages. 1986.

Vol. 5: **Mesenchymal-Epithelial Interactions in Neural Development.** Edited by J. R. Wolff, J. Sievers, and M. Berry. 428 pages. 1987.

Vol. 6: **Molecular Mechanisms of Desensitization to Signal Molecules.** Edited by T M. Konijn, P J. M. Van Haastert, H. Van der Starre, H. Van der Wel, and M.D. Houslay. 336 pages. 1987.

Vol. 7: **Gangliosides and Modulation of Neuronal Functions.** Edited by H. Rahmann. 647 pages. 1987.

Vol. 8: **Molecular and Cellular Aspects of Erythropoietin and Erythropoiesis.** Edited by I.N. Rich. 460 pages. 1987.

Vol. 9: **Modification of Cell to Cell Signals During Normal and Pathological Aging.** Edited by S. Govoni and F. Battaini. 297 pages. 1987.

Vol. 10: **Plant Hormone Receptors.** Edited by D. Klämbt. 319 pages. 1987.

Vol. 11: **Host-Parasite Cellular and Molecular Interactions in Protozoal Infections.** Edited by K.-P. Chang and D. Snary. 425 pages. 1987.

Vol. 12: **The Cell Surface in Signal Transduction.** Edited by E. Wagner, H. Greppin, and B. Millet. 243 pages. 1987.

Vol. 13: **Toxicology of Pesticides: Experimental, Clinical and Regulatory Perspectives.** Edited by L.G. Costa, C.L. Galli, and S.D. Murphy. 320 pages. 1987.

Vol. 14: **Genetics of Translation. New Approaches.** Edited by M.F. Tuite, M. Picard, and M. Bolotin-Fukuhara. 524 pages. 1988.

Vol. 15: **Photosensitisation. Molecular, Cellular and Medical Aspects.** Edited by G. Moreno, R. H. Pottier, and T. G. Truscott. 521 pages. 1988.

Vol. 16: **Membrane Biogenesis.** Edited by J.A.F Op den Kamp. 477 pages. 1988.

Vol. 17: **Cell to Cell Signals in Plant, Animal and Microbial Symbiosis.** Edited by S. Scannerini, D. Smith, P. Bonfante-Fasolo, and V. Gianinazzi-Pearson. 414 pages. 1988.

Vol. 18: **Plant Cell Biotechnology.** Edited by M.S.S. Pais, F. Mavituna, and J. M. Novais. 500 pages. 1988.

Vol. 19: **Modulation of Synaptic Transmission and Plasticity in Nervous Systems.** Edited by G. Hertting and H.-C. Spatz. 457 pages. 1988.

Vol. 20: **Amino Acid Availability and Brain Function in Health and Disease.** Edited by G. Huether. 487 pages. 1988.

NATO ASI Series H

Vol. 21: **Cellular and Molecular Basis of Synaptic Transmission.**
Edited by H. Zimmermann. 547 pages. 1988.

Vol. 22: **Neural Development and Regeneration. Cellular and Molecular Aspects.**
Edited by A. Gorio, J. R. Perez-Polo, J. de Vellis, and B. Haber. 711 pages. 1988.

Vol. 23: **The Semiotics of Cellular Communication in the Immune System.**
Edited by E.E. Sercarz, F. Celada, N.A. Mitchison, and T. Tada. 326 pages. 1988.

Vol. 24: **Bacteria, Complement and the Phagocytic Cell.**
Edited by F. C. Cabello und C. Pruzzo. 372 pages. 1988.

Vol. 25: **Nicotinic Acetylcholine Receptors in the Nervous System.**
Edited by F. Clementi, C. Gotti, and E. Sher. 424 pages. 1988.

Vol. 26: **Cell to Cell Signals in Mammalian Development.**
Edited by S.W. de Laat, J.G. Bluemink, and C.L. Mummery. 322 pages. 1989.

Vol. 27: **Phytotoxins and Plant Pathogenesis.**
Edited by A. Graniti, R. D. Durbin, and A. Ballio. 508 pages. 1989.

Vol. 28: **Vascular Wilt Diseases of Plants. Basic Studies and Control.**
Edited by E. C. Tjamos and C. H. Beckman. 590 pages. 1989.

Vol. 29: **Receptors, Membrane Transport and Signal Transduction.**
Edited by A. E. Evangelopoulos, J. P. Changeux, L. Packer, T. G. Sotiroudis, and K.W.A. Wirtz. 387 pages. 1989.

Vol. 30: **Effects of Mineral Dusts on Cells.**
Edited by B.T. Mossman and R.O. Begin. 470 pages. 1989.

Vol. 31: **Neurobiology of the Inner Retina.**
Edited by R. Weiler and N.N. Osborne. 529 pages. 1989.

Vol. 32: **Molecular Biology of Neuroreceptors and Ion Channels.**
Edited by A. Maelicke. 675 pages. 1989.

Vol. 33: **Regulatory Mechanisms of Neuron to Vessel Communication in Brain.**
Edited by F. Battaini, S. Govoni, M.S. Magnoni, and M. Trabucchi. 416 pages. 1989.

Vol. 34: **Vectors asTools for the Study of Normal and Abnormal Growth and Differentiation.**
Edited by H. Lother, R. Dernick, and W. Ostertag. 477 pages. 1989.

Vol. 35: **Cell Separation in Plants: Physiology, Biochemistry and Molecular Biology.** Edited by D. J. Osborne and M. B. Jackson. 449 pages. 1989.

Vol. 36: **Signal Molecules in Plants and Plant-Microbe Interactions.**
Edited by B.J.J. Lugtenberg. 425 pages. 1989.

Vol. 37: **Tin-Based Antitumour Drugs.** Edited by M. Gielen. 226 pages. 1990.

Vol. 38: **The Molecular Biology of Autoimmune Disease.**
Edited by A.G. Demaine, J-P. Banga, and A.M. McGregor. 404 pages. 1990.

NATO ASI Series H

Vol. 39: **Chemosensory Information Processing.**
Edited by D. Schild. 403 pages. 1990.

Vol. 40: **Dynamics and Biogenesis of Membranes.**
Edited by J. A. F. Op den Kamp. 367 pages. 1990.

Vol. 41: **Recognition and Response in Plant-Virus Interactions.**
Edited by R. S. S. Fraser. 467 pages. 1990.

Vol. 42: **Biomechanics of Active Movement and Deformation of Cells.**
Edited by N. Akkas. 524 pages. 1990.

Vol. 43: **Cellular and Molecular Biology of Myelination.**
Edited by G. Jeserich, H. H. Althaus, and T. V. Waehneldt. 565 pages. 1990.

Vol. 44: **Activation and Desensitization of Transducing Pathways.**
Edited by T. M. Konijn, M. D. Houslay, and P. J. M. Van Haastert. 336 pages. 1990.

Vol. 45: **Mechanism of Fertilization: Plants to Humans.**
Edited by B. Dale. 710 pages. 1990.

Vol .46: **Parallels in Cell to Cell Junctions in Plants and Animals.**
Edited by A. W Robards, W. J . Lucas, J . D. Pitts, H . J . Jongsma, and D. C. Spray. 296 pages. 1990.

Vol. 47: **Signal Perception and Transduction in Higher Plants.**
Edited by R. Ranjeva and A. M. Boudet. 357 pages. 1990.

Vol. 48: **Calcium Transport and Intracellular Calcium Homeostasis.**
Edited by D. Pansu and F. Bronner. 456 pages. 1990.

Vol. 49: **Post-Transcriptional Control of Gene Expression.**
Edited by J. E. G. McCarthy and M. F. Tuite. 671 pages. 1990.

Vol. 50: **Phytochrome Properties and Biological Action.**
Edited by B. Thomas and C. B. Johnson. 337 pages. 1991.

Vol. 51: **Cell to Cell Signals in Plants and Animals.**
Edited by V. Neuhoff and J. Friend. 404 pages. 1991.

Vol. 52: **Biological Signal Transduction.**
Edited by E. M . Ross and K . W. A. Wirtz. 560 pages. 1991.

Vol. 53: **Fungal Cell Wall and Immune Response.**
Edited by J. P. Latge and D. Boucias. 472 pages. 1991.

Vol. 54: **The Early Effects of Radiation on DNA.**
Edited by E. M. Fielden and P. O'Neill. 448 pages. 1991.

Vol. 55: **The Translational Apparatus of Photosynthetic Organelles.**
Edited by R. Mache, E. Stutz, and A. R. Subramanian. 260 pages. 1991.

Vol. 56: **Cellular Regulation by Protein Phosphorylation.**
Edited by L. M. G. Heilmeyer, Jr. 520 pages. 1991.

NATO ASI Series H

NATO ASI Series H

Vol. 74: **Molecular Mechanisms of Membrane Traffic.**
Edited by D. J. Morré, K. E. Howell, and J. J. M. Bergeron. 429 pages. 1993.

Vol. 75: **Cancer Therapy. Differentiation, Immunomodulation and Angiogenesis.**
Edited by N. D'Alessandro, E. Mihich, L. Rausa, H. Tapiero, and T. R. Tritton. 299 pages. 1993.

Vol. 76: **Tyrosine Phosphorylation/Dephosphorylation and Downstream Signalling.**
Edited by L. M. G. Heilmeyer Jr. 388 pages. 1993.

Vol. 77: **Ataxia-Telangiectasia.** Edited by R. A. Gatti, R. B. Painter. 306 pages. 1993.

Vol. 78: **Toxoplasmosis.** Edited by J. E. Smith. 272 pages. 1993.

Vol. 79: **Cellular Mechanisms of Sensory Processing. The Somatosensory System.**
Edited by L. Urban. 514 pages. 1994.

Vol. 80: **Autoimmunity: Experimental Aspects.**
Edited by M. Zouali. 318 pages. 1994.

Vol. 81: **Plant Molecular Biology. Molecular Genetic Analysis of Plant Development and Metabolism.**
Edited by G. Coruzzi, P. Puigdomènech. 579 pages. 1994.

Vol. 82: **Biological Membranes: Structure, Biogenesis and Dynamics.**
Edited by Jos A. F. Op den Kamp. 367 pages. 1994.

Vol. 83: **Molecular Biology of Mitochondrial Transport Systems.**
Edited by M. Forte, M. Colombini. 420 pages. 1994.

Vol. 84: **Biomechanics of Active Movement and Division of Cells.**
Edited by N. Akkaş. 587 pages. 1994.

Vol. 85: **Cellular and Molecular Effects of Mineral and Synthetic Dusts and Fibres.**
Edited by J. M. G. Davis, M.-C. Jaurand. 448 pages. 1994.

Vol. 86: **Biochemical and Cellular Mechanisms of Stress Tolerance in Plants.**
Edited by J. H. Cherry. 616 pages. 1994.

Vol. 87: **NMR of Biological Macromolecules.**
Edited by C. I. Stassinopoulou. 616 pages. 1994.

Vol. 88: **Liver Carcinogenesis. The Molecular Pathways.**
Edited by G. G. Skouteris. 502 pages. 1994.

Vol. 89: **Molecular and Cellular Mechanisms of H^+ Transport.**
Edited by B. H. Hirst. 504 pages. 1994.

Vol. 90: **Molecular Aspects of Oxidative Drug Metabolizing Enzymes: Their Significance in Environmental Toxicology, Chemical Carcinogenesis and Health.**
Edited by E. Arınç, J. B. Schenkman, E. Hodgson. 623 pages. 1994.

NATO ASI Series H

Vol. 91: **Trafficking of Intracellular Membranes: From Molecular Sorting to Membrane Fusion.**
Edited by M. C. Pedroso de Lima, N. Düzgüneş, D. Hoekstra. 371 pages. 1995.

Vol. 92: **Signalling Mechanisms – from Transcription Factors to Oxidative Stress.**
Edited by L. Packer, K. Wirtz. 466 pages. 1995.

Vol. 93: **Modulation of Cellular Responses in Toxicity.**
Edited by C. L. Galli, A. M. Goldberg, M. Marinovich. 379 pages. 1995.

Vol. 94: **Gene Technology.**
Edited by A. R. Zander, W. Ostertag, B. Afanasiev, F. Grosveld. 540 pages. 1996.

Vol. 95: **Flow and Image Cytometry.**
Edited by Alain Jacquemin-Sablon. 241 pages. 1996.

MIX
Papier aus verantwortungsvollen Quellen
Paper from responsible sources
FSC® C105338

If you have any concerns about our products,
you can contact us on
ProductSafety@springernature.com

In case Publisher is established outside the EU,
the EU authorized representative is:
Springer Nature Customer Service Center GmbH
Europaplatz 3, 69115 Heidelberg, Germany

Printed by Libri Plureos GmbH
in Hamburg, Germany

MIX
Papier aus verantwortungsvollen Quellen
Paper from responsible sources
FSC® C105338

If you have any concerns about our products,
you can contact us on
ProductSafety@springernature.com

In case Publisher is established outside the EU,
the EU authorized representative is:
Springer Nature Customer Service Center GmbH
Europaplatz 3, 69115 Heidelberg, Germany

Printed by Libri Plureos GmbH
in Hamburg, Germany